THIS WAS MINING

In the West

THIS WAS MINING

In the West

David W. Pearson

77 Lower Valley Road, Atglen, PA 19310

ACKNOWLEDGEMENTS

The author needs to acknowledge the contribution of Ron Bommarito in the compiling of this work. Virtually all of the photos which were used in this book were taken out of his collection, which is quite frankly the equal to or better than most museums. Most of the photos seen in this book have never before been published in book form. In his professional capacity as an antique dealer Mr. Bommarito has traveled all over the States of Nevada and California since before he was old enough to drive and is a source of much history that is simply unavailable from other published sources. He has been most cooperative in providing the author with whatever he asked and assisting whenever he could in developing information sources. The author also wishes to cite the help provided by the Carson City Library and their extensive collection of material on Nevada mines and history. Much of the bibliographic material was provided by the Carson City Library but several sources came out of the author's personal book collection and from the book collection of Mr. Bommarito.

The author also wishes to kindly acknowledge the forebearance of his wife Gerry, who works for the Carson City Library, and the efforts of a worn out old Pentax camera in the copying of many of these photos.

Printed in The United States.
ISBN: 0-88740-933-4

We are interested in hearing from authors with book ideas on related topics.

Published by Schiffer Publishing Ltd.
77 Lower Valley Road
Atglen, PA 19310
Please write for a free catalog.
This book may be purchased from the publisher.
Please include $2.95 postage.
Try your bookstore first.

Library of Congress Cataloging-in-Publication Data

Pearson, Dave, 1941-
This was mining in the west/Dave Pearson.
p. c.m.
Includes bibliographical references.
ISBN 0-88740-933-4 (paper)
1. Gold mines and mining--West (U.S.)--History.
2. Silver mines and mining--West (U.S.)--History.
3. Copper mines and mining--West (U.S.)--History. I. Title.
TN423.A5P43 1996
338.2'742'0978--dc20 95-20554
CIP

CONTENTS

MINING SITES IN THE AMERICAN WEST
WASHINGTON
MONTANA
NORTH DAKOTA
OREGON
IDAHO
SOUTH DAKOTA
WYOMING
NEVADA
NEBRASKA
UTAH
COLORADO
KANSAS
CALIFORNIA
ARIZONA
NEW MEXICO
OKLAHOMA
TEXAS
CALIFORNIA
1. Amador County
2. Bodie
3. Mariposa County
4. Mother Lode District
5. San Francisco
6. Sacramento
7. Sutter's Mill
8. Cerro Gordo
NEVADA
9. Aurora
10. Austin
11. Carson River
12. Ely
13. Eureka
14. Genoa
15. Gold Hill
16. Goldfield
17. Hamilton
18. Lake of the Sky, Lake Tahoe
19. Las Vegas
20. Ruth
21. Silver City
22. Tonopah
23. Tybo
24. Virginia City (The Comstock)
25. Ward
COLORADO
26. Denver
27. Cripple Creek District
28. Pike's Peak
29. San Juan Mountains
30. Telluride
SOUTH DAKOTA
31. Black Hills
32. Deadwood
33. The Homestake Mine
MONTANA
34. Alder Gulch
35. The Anaconda
36. Butte
37. Montana Bar
OREGON
38. Canyon City
39. Sumpter
40. Virtue Mining District
WASHINGTON
41. Ft. Vancouver
42. Blewett
43. The Eureka District
44. Fort Colville
45. Liberty
46. The Republic Mine
47. Ruby
48. Salmon
WYOMING
49. Atlantic City
50. South Pass City
UTAH
51. Bingham Canyon
52. Corinne
53. Mammoth
54. Salt Lake City
55. Tintic
IDAHO
56. Coeur d'Alenes
ARIZONA
57. Gila City
58. Tombstone
The Emigrant Trail

INTRODUCTION

Working a high ledge placer claim the hard way. The shaft shown behind the miner was sunk to paydirt. Then the gravel was hauled out and sluiced through the rocker box, with what little water could be hauled up the hill from the creek. This was sometimes the only way to reach gold found in higher locations. Collection of Ron Bonmarito.

Mining in the western United States was fundamentally different from mining in the rest of the country and most of the known world. Although it is now known that Spanish settlers and the Indians before them had been mining precious metals for centuries, it is commonly accepted that mining in this region started during the late 1840s. Mining as discussed in this book is the search for precious metals—first gold and silver, and later copper.

Little was known at the time of what existed west of the Mississippi River other than vast open spaces, enormous mountain ranges, hostile Indians, great herds of bison, beaver in every stream, and little else. Indeed, much of the area was simply pictured on early maps as the Great American Desert, even though it was land that did not belong to the United States. This was the political period of Manifest Destiny, when Americans widely believed that the United States had a godgiven right to take whatever it wished.

The earliest mining pioneers went West to look for gold or silver. Little else interested them even when they stumbled across it. Copper was found in great quantities, for instance, but its low value and problems associated with refining it meant that this ore was frequently overlooked or discarded unless it contained high percentages of silver.

It should also be noted that the ultimate buyer of gold and silver was the government, which set the purchase price. During this period the United States was on a hard money standard; currency was gold or silver coinage. Virtually everyone scoffed at paper money, since the banking industry as it then existed was very unstable. The law allowed paper money to be printed by almost any organization with a printing press, and all too frequently paper money was not worth the price of the paper it was printed on—literally! There are several recorded instances in which bandits looking for gold turned down paper money.

In its raw form, as it was taken from placer and lode mines, gold could be and was used for barter purposes. While raw gold is normally not pure, it was close: 90-95% pure (about 22-23 carat), thus considered acceptable for barter at face value in mining towns. Pure gold, 24 carat, is too soft for jewelry, coinage, or other normal uses, since it is easily bent and damaged. Gold coins vary from as little as 50% to about 90% purity.

Raw gold that is panned out of a stream has a distinctive type of coloring that once seen is always recognizable. Old-timers from a particular mining district have been known to determine the exact location of the mine that produced a particular nugget strictly from the color and roughness of the gold. On a more scientific basis, metalurgists are able to tell exactly where gold comes from by doing detailed analyses to determine the trace elements, usually silver and copper, that exist in samples. By doing this sort of sampling it is possible to tell if a mine has been salted or if a piece of gold was stolen from a particular miner.

In the earliest days of mining in the west no such niceties existed: Gold was Gold! The United States was a growing country and it needed a steady, continually growing supply of new hard metal. Until the first California rush, much of this hard metal came from off-shore in the form of British, French and Spanish coinage. What did come out of the eastern U.S. came primarily from small-scale finds in a few southeastern states, a byproduct from the mining of lead and copper, two metals in wide daily use. Until the first discovery at Sutter's Mill, the United States had no good sites for mining either gold or silver.

Sutter's Mill was a water-powered saw mill east of Sacramento, in the Sierra Nevada foothills. There, in December of 1848, gold was found, laying about in piles. Word got out in 1849 about the Sutter's Mill gold, and the 49ers gold rush started, to continue in stages for many years after. Rumors had it that California gold was 'ready for the picking'.

Actually, in the earliest days of the California rush of '49 the easiest pickin's were almost as easy as that. The simplest method of mining was 'placer mining'. In placer mining, one or two laborers (or more, if you had a few trustworthy friends) built a simple long wooden box, called a 'sluice box', with crosswise slats called 'riffles' nailed into the bottom of the trough. Ore-bearing dirt was shoveled into the box while a constant stream of water ran through it. The gold, which was heavy, drifted to the bottom of the stream and was caught in the rifles, while the lighter gravel or sand was washed away by the volumes of water. Quicksilver (mercury) could be put into the riffles to combine with the gold, and make its recovery easier, since quicksilver could catch the smaller gold dust particles that would have been carried off by the water otherwise.

The placer mining method was (and still is) fairly easy to use, allowing a relatively poor but strong individual to work a claim by himself once it was set up. It was not especially efficient from a recovery standpoint, missing much of the fine-grade gold. If the miner was lucky enough to find a hot spot, he could produce a sizable fortune in as much time as it took him to shovel down to bedrock; since gold is heavier than most metals, it tends to work its way down to the lowest point. All of the early tailings (waste rocks) were reworked in later years and substantial amounts of gold were recovered by more efficient dredges and even by Chinese laborers.

This was how the earliest gold was found in California, Nevada, and many of the other western states. The image of the single wizened prospector with his pack mule, shovel, gold pan, and trusty Colt revolver is forever etched into the history of the West. However, we know today that the reality of mining was much different. That is what this book is all about.

CALIFORNIA

The earliest gold finds in California were in the southern parts of the state, and were worked by Indian or Mexican laborers using slow and labor-intensive methods. In general, they had low yields—the gold was sparse, and often came in very fine particles, almost dust. However, after the first big find in Sutter's Mill in 1848, everything changed. The gold found at Sutter's Mill was in the form of fairly large nuggets—lots of them—which the mill workers had been able to pick up by hand from the wooden mill-race of the saw mill. Word about these gold nuggets spread like the plague, as did prospectors looking for more pay dirt.

These early strikes were all placer finds, and the miners quickly concluded that there must be a mother lode of gold ore. Gold found in the size and quantity of that found at Sutter's could only have come from a lode—or vein—somewhere upstream from the mill. Gold can be transported by water quite a distance, but the farther it travels the rounder the nuggets become, from being tumbled across the rocks at the bottom of the stream bed. When miners find large, rough nuggets like those at Sutter's, and no fines (small flakes or dust), they have a good idea that they are close to the source—the "Mother Lode."

So inspired, it didn't take the early miners long to discover veins, and then to stake hardrock subsurface claims. The search eventually led to the defining of the true California 'mother lode' district in the foothills of the Sierra Nevada mountain range. This district covers an area about 120 miles from north to south and 40 miles from east to west. This mother lode is now known to be a series of dikes or intrusions ('veins') into the hard rocks which comprise these rugged, geologically active, difficult-to-traverse mountains.

The earliest lode claims in California's mother lode district were in Mariposa County in 1849 and in Amador County in 1850. These were called "gold quartz claims" (a term synonymous with subsurface hardrock claims or lode claims), and by local miner's rule could measure no more than 120 feet by 60 feet per individual. In the early days there was no federal law governing the miners; California was not even a state yet, let alone one with a well-developed set of laws. Mining Law was determined by a consensus of local miners, varying widely from place to place and from time to time. In actuality, it was total confusion.

Getting down to the bottom of bedrock in the Golden Gate placer claim, Feather River, Butte City, California about 1891. Note the Chinese laborers. From a photograph by Carleton Watkins, an early photographer of mines and cities in the American West. His photographs are of a very high quality. Collection of Ron Bonmarito.

The mines in Granite State, Pleasant Ridge, and Spring Hill were all early producers of quartz ore in Mariposa and Amador Counties. The major towns that were affected by quartz mining were Grass Valley and Nevada City, both in Amador County and both still healthy towns

In placer mining, there is one essential ingredient: water. At times the Sierras provided this element in overabundance, but at others it provided none at all. Hardrock mining, on the other hand, needed an abundance of a different sort: capital, in the form of money. Finances were required to purchase mountains of equipment, and, most of all, to pay the miners' wages. Because the enormous expenses could seldom be borne by one man, companies were formed.

As the second, third, and fourth rushes of miners came to California through the 1850s, they discovered not gold itself but the hard, dirty work of subsurface hardrock mining for wages. Working in these mines paid well because it was difficult, dangerous, unglamorous work for wages, not for gold itself. The early subsurface miners toiled with hand tools to wrest ore from the unforgiving, unyielding mountains. Accidents were frequent, maiming, and commonly fatal.

When a vein or lode was located, it was necessary to follow it in whatever direction it went. This usually meant tunneling. Generally, gold-bearing white quartz was the carrier ore, and since it stood out visually from the surrounding rock it could be followed by the naked eye. This carrier rock was first surrounded by holes drilled by hand, either 'single jack' or 'double jack'. Single jack drilling was done by one man with a set of drills and a hammer; double jack drilling was done by a team of two men alternately holding and then hitting the drill bit. Whichever method was used, it was strenuous, dangerous work. Once the holes were drilled to the correct depth (usually about four feet deep) the powder man was called for. He set his charges, using coarse grade black powder in small packages. This was tamped into the holes with wooden sticks to avoid an errant spark, and then it was fused with the correct length of fuse. All the holes were set in the same way, to create an explosion that would break loose a section of rock and ore all at once.

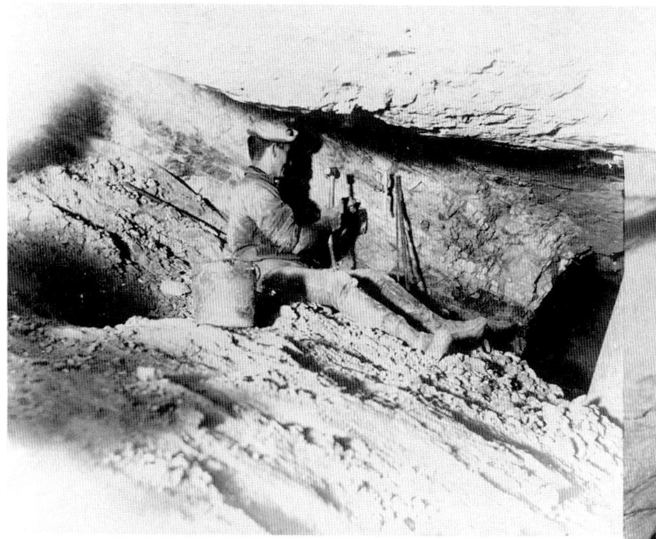

'Single jacking' the ore face. When the miner came to work he stayed there until it was time go home. He had his candlestick for light and a bucket for food and water. Collection of Ron Bonmarito.

Possibly a staged photo, this is a good illustration of the 'double-jacking' method of drilling rock. The two men at right are doing just that, with one man holding the steel and the other hitting it with a sledgehammer. Double-jacking was faster and could drill deeper but required two men who had confidence in each other and lots of experience working with each other. Note the pile of drill steel at left. From the Empire Mine. Collection of Ron Bonmarito.

Powder men were seasoned professionals—well paid, highly skilled at their trade, and held in high esteem by their co-workers. Once the shout of "Fire in the hole!" was heard, their ability was often the difference between life and death for everybody in the stinking, smoke-filled hole. It was not until the late 1860s, when dynamite became available, that their charges became a bit easier to set. Even then the occasional short fuse lead to death and dismemberment among miners and powder men around the world.

Dynamite in its various forms has the annoying tendency of "leaching out" the nitroglycerin used in its manufacture. Nitroglycerin in solid form is fairly stable, but in liquid form it is very unpredictable, especially if exposed to heat. Sticks of dynamite contain a stabilizer that makes the nitroglycerin jell into a semi-solid state, which is then wrapped in paper. When the nitroglycerin "leaches out," it falls out of suspension in the jell, and forms in actual liquid droplets on the surface of the paper. This is a very dangerous situation! It is particularly bad if the explosive is subjected to heat over a prolonged period of time; it then becomes so touchy that merely picking up a stick can cause an explosion. The resulting powderhouse explosions were frequent occurences in early mining towns.

The powder man lighting the fuses. Quite a remarkable photograph, if he was actually lighting a charge. This was taken in the Empire Mine in the 1890s. The quartz vein being worked is just to the left of the man's hand and above his head. Collection of Ron Bonmarito.

Three men working the ore face with single hammers and drills. They are probably picking out loose ore after a shot. Collection of Ron Bonmarito.

An early photograph of the chlorination works and arrastras in Nevada City, California. These arrastras are all water-powered, using water from the flumes to both rotate the grind stones and flush the milling areas. The large pole in the center is a mast for operating a block and tackle to move ore into the arrastras. Collection of Ron Bonmarito.

Once ore was removed from the vein, it had to be moved to the surface. This was not a easy task. The first job was to load the ore into ore carts down inside the mine. Many men, mules, burros, and horses were used to move carts filled with rock to the chutes, where the ore was moved to another track system. This system moved the ore to the next step, the lift station. In early mines the lift stations consisted of large iron buckets standing three or four feet deep, each big enough to hold a man. These buckets were filled with ore, and then hauled to the surface. There, the ore was dumped into hoppers or into carts used to move the ore and waste rock around on the surface.

Different mines could use different systems. Some mines did not transfer the ore to buckets before lifting it to the surface; instead they lifted the full carts themselves. This eliminated some of the effort of handling the heavy ore.

Once extracted from the mine, the ore had to be processed. This involved crushing the ore to get at the usually chemically free gold. Initially, this was accomplished by using a tool known as an 'arrastra', which had been introduced by Mexicans. An arrastra—much like a mill wheel in operation—was a flat, saucer-shaped rock with a pivot pole in its center. An arrastra could be quite large, often with a base 20' in diameter. A large, flat stone was dragged around the pivot pole by a mule or horse. The drag stone needed to be of relatively hard rock, and was commonly as large as 4' by 6' by 1'. The gold-bearing ore was introduced into the path of the flat drag stone, and was eventually crushed as the drag stone was pulled over it. The gold, sand, and residue could then be removed for panning or smelting.

The arrastra continued to be used as a device for reducing ore (in some areas, into the 1870s) largely because it was inexpensive to build. Still, the arrastra method was slow, inefficient, and generated a lot of waste rock that had to be removed regularly. The process remained profitable only so long as high-grade ore was found. When lower grade ore was used, the expenses of the arrastra method outweighed the value of the gold retrieved, and mines abandoned the dragstone in favor of wooden frame stamp mills.

Wooden frame stamp mills also wore out quickly. Since the frames used to hold the stamp mills were entirely made of wood, the combination of a lot of weight, moving rocks, and moving water quickly wore the components out. Wooden frame stamp mills were eventually replaced by all-iron stamp mills. The iron versions were longer-lasting, but more expensive to build. They usually came in ranks of five.

No matter what their construction, stamp mills all had two things in common: they were noisy, and they had to be run by a separate power source. In its simplest form, this power source was a water wheel. The wheel rotated a shaft, to which many cams were attached. These cams lifted stamps up, and then let them drop the twelve to eighteen inches necessary to crush the ore. The crushing head was cast iron, and quite heavy. The ore was fed into one side by gravity, and the fines (powdered gold and small gold particles) were moved out the other by a water flow. If a large hunk of ore was too big to fit into the stamp mill, a solid blow with a sledge hammer was the solution. Eventually circular crushers were introduced; these broke the ore into managable fragments before it was smashed by the stamp mills. In response to the mining industry's need for tools like these, a number of iron foundries were established in the mother lode district.

MINERS' TOOLS

A word is in order here about miners and their tools. As anyone who has every been in a mine or cave can attest, without a light source the darkness is absolute. Initially, candles were the usual form of illumination for miners. Miners' candlesticks were hand-forged by blacksmiths, some virtually works of art. Various types of holders were developed to keep the candles out of the way, but still near enough to work by; they were made to be hammered into wood or rock, or hung over a convenient beam. They are much-sought-after today and bring premium prices.

In England, coal miners used a special type of candlestick lamp developed especially for them to solve the problem of methane gas. Methane (the primary gas used in home gas systems) is highly explosive; indeed, it is the cause of most mine explosions. The English miners used candlesticks specially rigged to protect them from methane explosions. The candlesticks were equipped with little brass devices that kept methane from reaching the open flame, by forming a barrier on which the gas actually burned off first—quite effective in practice. Miners in the western U.S. seldom found use for lamps like these, however. Because methane is a product of decomposing organic matter, it is frequently encountered in coal mines, but not often in gold mines.

Oil lamps were the next stage, but these devices were little better than candles, and needed a supply of kerosene or some other combustible oil. Eventually, the carbide lamp was developed; it was fairly reliable and gave a brighter light than other sources, although it needed constant adjustment and care. Now, of course, miners use battery-powered lights.

The shift on the way down to work. Everybody has a candlestick! The old gent in front was probably a foreman. Notice the young boy at right wearing fancy knickers. Collection of Ron Bonmarito.

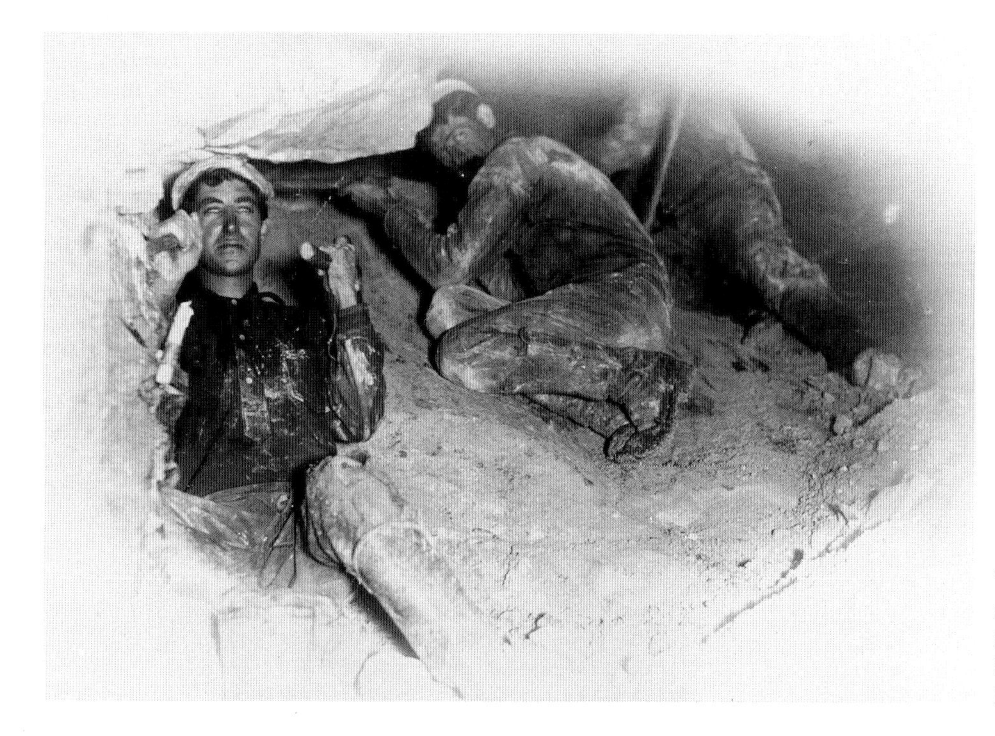

Working the ore face by hand. Note the candle holder in near left. The two men at rear appear to be double-jack drilling. Empire Mine. Collection of Ron Bonmarito.

MINING CLOTHING

Miners wore whatever clothing worked at the moment. Oilskins were common, as were parafin-impregnated trousers and coats. If the mine was cold, and many were, several layers of clothes were worn so the miner could adjust his coverings as needed. In the Comstock Lode mines in Virginia City, Nevada, the problem was heat. Miners there wore heavy-duty clothes on the way down, but (as can be seen in these photographs) they quickly stripped to the waist as they encountered horrendous temperatures inside the mines.

Miners were quick to discover that cuffs on pants were handy places for hiding a few pieces of high grade ore to take home with them at the expense of the company. For a long time, companies more or less overlooked this practice, but eventually the owners decided that they were being robbed blind by the workers. They not only banned the practice of hiding gold in cuffs, but went so far as to ban cuffs on clothes altogether. Eventually they built separate changing rooms and required the miners to change clothes before going home. Some later mines had showers, the drains of which were designed to capture any dirt washed off the men, and then put it through a reclamation process.

Early miners tried using the rocker box and a trickle of water, from the left. Note the clothes of this early miner in California. Collection of Ron Bonmarito.

MORE MINING DEVELOPMENTS

Shallow placers that could be worked by hand easily were rapidly being depleted in California by 1855, a mere six years after the first strikes. Underground quartz mines were expected to produce gold, but the high cost of milling hardrock ores was a deterrant to mine owners. In general the cost of milling increased from roughly $7.50 per ton to as high as $15.00 per ton in the 1880s. Add to this the cost of labor: about $3.00 per shift for each man who lived off the mine premises, and most did. With expenses like these, the ore had to be of good quality for the mine to make a profit.

When it became possible to produce electricity at the mines this innovation was quickly adopted. Similarly, compressed air drills were used when they became available. Subsequently, some of the underground quartz mines followed ore veins that went quite deep. By the time of the First World War they had reached depths of 3000 to 4000 feet.[1]

Almost concurrent with the development of the deep quartz mine in California was the initial use of Little Monitor water "Giants" to remove overburden (waste rocks and dirt). The Giants required massive amounts of water to wash away hillsides and get at the gold-bearing paystreak near the bedrock, a process called "hydraulic mining." While the Monitors were efficient at removing the overlying dirt of placer deposits they were also terribly destructive to the countryside. In addition, the enormous amounts of water they used sometimes had to be moved very long distances through flumes or pipelines.

During the second decade of the California rush—no longer a rush, really—hundreds of canals, ditches, flumes, and tunnels were built to bring water downhill to feed the Giants. This must have involved quite a large financial investment, but there is no single source that agrees on a figure. It is likely that approximately $30 million was invested between 1860 and 1880, to build over 7000 miles of ditches (estimated) and an unknown number of pipes.

Hydraulic mining. After the Monitors wash off all the overburden, the crevices in the bedrock must be examined in detail for any nuggets missed during the sluicing. The two men on the left are using whiskbrooms and picks to clean up the rock. Collection of Ron Bonmarito.

A "Little Monitor" water Giant in Auburn, California. Date unknown but probably as early as 1870. Collection of Ron Bonmarito.

A very early attempt at hydraulic mining in the Auburn, California area. Compare the size of the nozzle and smallish stream of water with those in the Malakoff mine shown in other photographs. The idea in hydraulic mining is to get as much water into the smallest area moving as fast as possible, but volume is the most important component. It took a while to figure this out. This photograph dates from the late 1860s or early 1870s. Collection of Ron Bonmarito.

Opposite page, top:
The Malakoff Mine in North Bloomfield, Nevada City, California about 1871. To get an idea of the scale, note that there are three men in the area of the nozzle: one to the left, one right next to the nozzle, and one to the right under the water arch. Collection of Ron Bonmarito.

Opposite page, bottom:
Overall view of the Malakoff Mine. Compare this with the other photograph of the same mine and it is possible to get some idea of the devastation caused by hydraulic mining in this era. Collection of Ron Bonmarito.

By design, hydraulic mining was a fairly labor-efficient process. Once the water was turned on it took only one or two nozzle men to keep the water running down a hillside and into the long prepared sluices, which extracted just about everything there was to be gotten out of the gravel. Additional workers were necessary only for relief and shift changes. Most of the men in this type of operation were employed in maintaining the water system when it was turned off and in the clean-up of the resulting concentrate in the sluice boxes.

While the efficiency of hydraulic mining was good news for California mining companies, it only made their business more destructive, since they were moving thousands of acres of hillside down to the productive central valleys below. The rich valley farmlands were being cultivated in the 1860s and 1870s, and the farmers did not take kindly to their lands being flooded every year because hydraulic mining debris choked the rivers. The court system got involved, and by 1884 most of the hydraulic mines had been shut down.

The mining companies were innovative, though, and began to mine gravel at the bottom of rivers and other wet areas that had been untouchable before. To do this, they used large captive floating dredges, which were moored to the banks. The dredges, mechanical monsters, were very good at mining placer gold, and were fine-tuned into highly efficient machines. Unfortunately, they too left a moonscape of tailings behind them. These dredges were one of the last methods of placer mining to be used in California, with some still operating during the 1930s. In some areas, the barren piles of rock from the dredges still prohibit the growth of anything but the hardiest of weeds.

Early on, the need for mercury to amalgamate gold and silver ores prompted miners to search for a local source of the metal. This is the New Almaden Mine in Central California, the biggest producer of mercury from cinnabar ore in the West. This photograph was taken around 1863 and is thought to be an original Watkins photograph. Collection of Ron Bonmarito.

Although the exact date of this photograph is unknown, it is probably about 1900. It shows the business end of a large floating dredge in Marysville, California. Collection of Ron Bonmarito.

The really ugly side of gold mining: refining in a smelter. This is a large smelter operation, probably in California. Note the two men in center for scale. Collection of Ron Bonmarito.

After the smelter got done, this is what they produced. Gold bars like this were quite heavy, in the neighborhood of 85 pounds or more and about 95% gold. Further refining would recapture the small amounts of silver and copper impurities left and reduce the gold to .999 fine, pure gold. Collection of Ron Bonmarito.

An 1879 photograph of Bodie, California; the date was determined by the absence of certain buildings. Take note of the barren nature of the mountains surrounding the town. The reason for the lack of trees is that the town is above the treeline at 9000 feet in elevation. Collection of Ron Bonmarito.

This photograph is undated but was taken in Bodie, California. Note the size of the man kneeling in the center. His name is believed to be George Moore and he is BIG! Collection of Ron Bonmarito.

A winter photograph of Bodie's Main Street. This picture was taken in 1923.

Another photograph of Bodie's Main Street. This picture was taken from the opposite direction (see previous photograph), at a slightly different time. Collection of Ron Bonmarito.

One of the storefronts of Bodie during the winter. No date appears on this photograph and the original is blurry. The women in the center appear to be Indians. Note the foot of the person on the snow bank at upper left. Collection of Ron Bonmarito.

Bodie in 1901. This was a large cyanide processing plant in use around that time. Collection of Ron Bonmarito.

On the back side of the mountain beside Bodie. Mining was actively carried on underground at this altitude. Mono Lake is in the background. Collection of Ron Bonmarito.

A tramway and ore cars for moving ore to the mill from the various mines on the mountain in Bodie. Collection of Ron Bonmarito.

Mill equipment in Bodie around 1910. This complicated affair is some sort of classifier or crusher. This photograph was taken by McClure, another early chronicler of the American West. Collection of Ron Bonmarito.

Not all the high altitude mining was done at Bodie. This was Lundy, California around 1900. This is the lower tunnel and the tram and mill. The original photograph is badly damaged. Collection of Ron Bonmarito.

The electric power plant in Lundy, California, built in 1892 although this photograph was taken in 1908. The steam turbine is at lower left while generators are at right rear. Collection of Ron Bonmarito.

The Lundy, California mine is at upper left rear. "X" marks the spot in the snow field. This photograph was taken by Hawes, an early photographer from Carson City, Nevada who took many pictures of the Eastern Sierra. Collection of Ron Bonmarito.

The rope tram for moving ore carts to Lundy from the mine. Photograph taken July 6th, 1893. Collection of Ron Bonmarito.

An early effort at hydraulic mining in Auburn, California. Probably mid-1870s. Collection of Ron Bonmarito.

Hydraulic mining in Auburn California. Collection of Ron Bonmarito.

CERRO GORDO

One final California mining project begs to be mentioned: Cerro Gordo, a frustrating affair that started during the 1850s and continued into the 1860s. Cerro Gordo has more in common with Nevada mining than with California mining, because it produced silver-lead bullion rather than gold. Located on the eastern side of Owens Valley, the various mines that made up Cerro Gordo were all located on one large mountain. They were at an elevation of almost 9000 feet above sea level, nearly a mile above the valley floor, with no source of water. Since Cerro Gordo miners were not looking for gold but for silver and lead (which do not require water to retrieve) the lack of water did not ruin the mines' prospects. The biggest problems at Cerro Gordo were matters of simple logistics: in the summer the area was brutally hot and very dry; in the winter, snow and cold weather was normal. This made living, let alone mining, a difficut job.

The ore mined at Cerro Gordo was very high-grade lead-silver ore, but it was also very complex with other trace minerals. Significantly, the mountain produced different ores in different locations. Some were much higher in silver than in lead; these were the first areas to be worked. The earliest efforts at smelting this ore used old Mexican methods—simply burning the rocks in a 'pancake stack' of firewood. This drove off most of the trace elements; however, the process was slow, the recovery of the metal afterwards was a tedious job, and the results still were not pure silver. Furthermore, wood was not in great supply in this arid region, and so had to be hauled in over long distances over the roughest roads imaginable. In addition to the difficulty of keeping a smelter stoked with firewood in Cerro Gordo, the smelters had a nasty habit of burning down when they were stoked. Cerro Gordo mining companies eventually gave up trying to smelt their ore on site, and instead contracted with outside smelters.

Because it was difficult to get the ore to an outside smelter and from there to market, true development at Cerro Gordo was slow. Through trial and error, the mine owners finally developed a furnace that could convert the complex ore to bullion. The ore was in fact more lead than silver, with traces of gold and zinc in it. When the bars of bullion (called "DORE" bars) were tested they were about eighty-three pounds each, with eighty of that being lead. Each bar contained several ounces of

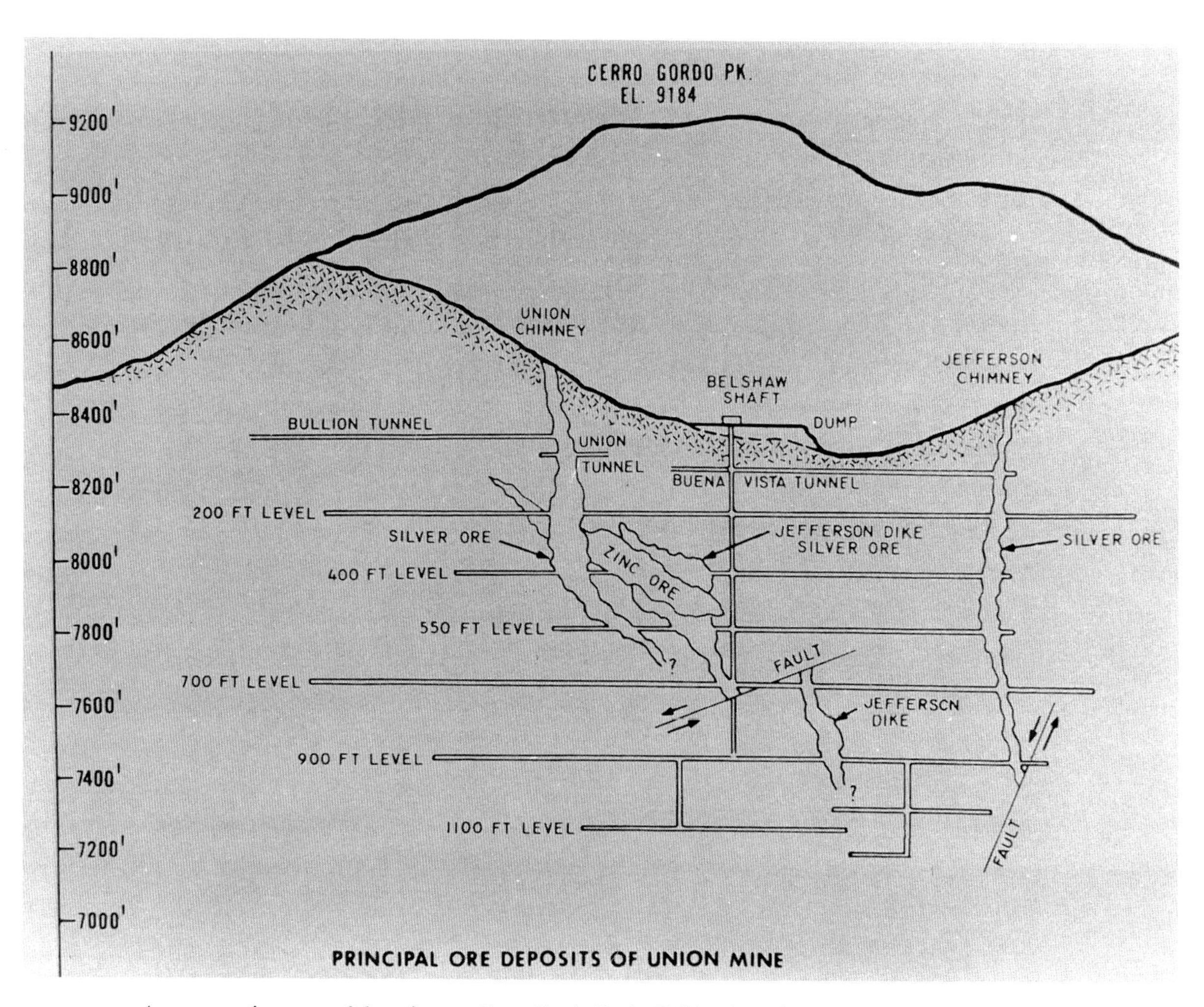

A cross-section map of the mines at Cerro Gordo Peak, California. Take note of the various ore bodies and the considerable faulting as well as the altitude. There were other smaller pocket mines on the surface of Cerro Gordo, but the Union Mine was the largest single mine and accounted for most of the production from this mountain. Author's Collection.

A mule train waiting to be loaded at the mill in Cerro Gordo, California. The original photograph is damaged. Collection of Ron Bonmarito.

silver, a trace of gold and some zinc, all metals that could be recovered by a proper smelter. The nearest smelter capable of handling this kind of ore when production started was in San Francisco. The owners contracted with a freight hauling firm to carry the bullion, seven tons at a time, first to Los Angeles and then by ship to San Francisco.

Luckily for the Cerro Gordo owners, there was a ready market for the lead in San Francisco: a major lead shot producer named Thomas Selby. Selby took every bit of lead they could deliver to him and handed them back silver and small amounts of gold. In this case, silver was the mining company's profit, but it was lead that paid the bills.

When the main on-site smelter burned in 1877, it spelled the end of Cerro Gordo. Workers deserted the mines en masse.

The Integral Mine in Trinity County, California. About 1890. Collection of Ron Bonmarito.

The Integral Mine in Trinity County, California, thought to have been a cinnabar mine. Collection of Ron Bonmarito.

The assay office of a Mr. S. A. Gilchrist in Sonora, California, one of the early mother lode towns and still in existance. Collection of Ron Bonmarito.

ASSAYING IN ALL ITS BRANCHES
CYANIDE, CHLORINATION
MILL TEST

S. A. GILCHRIST

MINING PROPERTIES
REPORTS ON MI
GOLD

Tuolumne Mining Bureau

Washington Street

Sonora, Cal.________________

Deposited by________________________________

	SILVER Ounces per Ton		SILVER Value per Ton At____per Oz.		GOLD Ounces per Ton		GOLD Value per Ton		VALUE OF GOLD AND SILVER PER TON		Per Cent of
	OUNCES	DECIMAL	DOLLARS	CENTS	OUNCES	DECIMAL	DOLLARS	CENTS	DOLLARS	CENTS	

Very respectfully,

________________________ Assaye

One of Gilchrist's assay report forms. Note there is no place for anything other than gold or silver. Nothing else mattered! Collection of Ron Bonmarito.

Greenwater, California around 1906. Greenwater was in the Death Valley area but is now thought to have been a scam, as no gold was ever found there. Collection of Ron Bonmarito.

The Greenwater Times "buildings." The town folded up and left almost as soon as it was started. Collection of Ron Bonmarito.

Single-jack drilling in California hard rock mine. Note the two men working the ore face at lower left and the candle holder stuck in the rock. Collection of Ron Bonmarito.

Opposite page:
A very early photograph of the shaft of the Empire Mine in Nevada City, California. Note the use of wooden rails for the iron-wheeled ore carts. Collection of Ron Bonmarito.

The mine hoist of the Empire Mine, showing an ore cart at right with a crew cart at left. The shaft of the Empire was inclined at about 30°, which followed the original vein. Collection of Ron Bonmarito.

Below:
What the miner faced when going to work. This was a high quality mine; note the professional stonework over the opening. Also note the steps in between the rails. If need arose, the miners could walk out, about 2500 feet. Collection of Ron Bonmarito.

'Mucking' the ore meant shoveling ore into the cart by hand. This was back-breaking work and the mucker was expected to load 16 tons (about 20 cartloads) of rock a shift. This photograph is unusual as it shows the irregular surfaces and open areas that were frequently encountered in deep hardrock mines. From the Empire Mine. Collection of Ron Bonmarito.

Major timbering and filling using waste rock. Note the height of tunnel. The area in front was probably a 'glory hole' or a large, very rich, concentrated deposit. Collection of Ron Bonmarito.

EMPIRE MINE

Dumping the ore carts down the chute to be collected and hauled to the surface. Empire Mine. Collection of Ron Bonmarito.

Working the ore face. These men are probably taking off ore after a shot. This photograph clearly shows the difference in rock between the quartz-bearing ore and the waste rock above. Empire Mine. Collection of Ron Bonmarito.

An early compressed air drill. The men hated them because they were noisy and hard to work with, but they greatly simplified the task of drilling and measurably speeded up production. Note that they are still using candlesticks for illumination. Probably Empire Mine. Collection of Ron Bonmarito.

Below:
The foreman's station. He had already paid his dues and got to supervise on the job. This photograph in the Empire Mine is from the 1890s and shows a bell system for relaying messages to the hoist man and a telephone. The bell was probably more reliable than the telephone. Also note the pile of drill steel on the gound waiting to be taken up for sharpening. Collection of Ron Bonmarito.

Ore-lift carts used solely for taking ore to the surface. They were dumped into ore-processing bins or tailings bins according to the instructions of the men in the hole. Collection of Ron Bonmarito.

Below:
When the ore was dumped into the hopper under the hoist, it was put first through a conical crusher to reduce it to a workable size before going to the stamp mills. Empire Mine. Collection of Ron Bonmarito.

Opposite page, top:
One of the ill-treated burros or mules used in the mines to haul ore cart trains to lift stations. These animals frequently lived out their lives underground. Life expectancy was probably not high for these animals. Empire Mine. Collection of Ron Bonmarito.

Opposite page, bottom:
A cart-loading station from an ore chute. The man on the boards opens the chute doors with the carts underneath. Empire Mine. Collection of Ron Bonmarito.

The bottom of the conical crusher. This type of crusher is still in use. Empire Mine. Collection of Ron Bonmarito.

Bottom:
The stamp mill, showing the metal trays or plates below the actual stamps. These plates or trays were coated with mercury to catch gold and silver as it was liberated in the milling process. Collection of Ron Bonmarito.

Top view of the stamp mill. This is a good detail of the cams, lifters, and rotary belts powering the whole operation. Empire Mine. Collection of Ron Bonmarito.

The lower pan works beneath the stamp mills. This was nothing but a riffle box system with quantities of mercury in it to hold the gold as it came through in the slime. Water is continually used to move the tailings through. This photograph shows it in actual operation. Empire Mine. Collection of Ron Bonmarito.

Clean-up time at the mill. Stamps in this mill are of a larger size. The work area in front of the men is sheathed with copper and coated with mercury. The resulting amalgam is removed using a wooden paddle that the one man is using. The amalgam is then put into...what else, a gold pan. Empire Mine. Collection of Ron Bonmarito.

Unwanted rock was brought to the surface and dumped into a hopper, then moved by cart to the tailing dump. Empire Mine. Collection of Ron Bonmarito.

Tailing piles got bigger and bigger until they covered an amazing amount of ground. The track was continually extended as needed to dump the rock. Empire Mine. Collection of Ron Bonmarito.

The Empire Mine hoist and tailing piles. Note the various 'tipples', or 'bunkers', for ore and waste. All the rocks in the center are tailings. Collection of Ron Bonmarito.

A view from the top of the hoist of the Empire Mine. The hoist is dumping an ore cart on the left track in lower right center. Collection of Ron Bonmarito.

The shift change. The tired crew goes home, each man with his candle holder in hand. Note the early light bulb at left. Empire Mine. Collection of Ron Bonmarito.

The Empire Mine hoist room. One of the really important people at the mine was the hoist man. He had to be a sober individual by nature. Communication with the lower sections of the mine was best done by a system of bells, in this case electric bells. Phones were unreliable and it was hard to hear in the din. The large dials indicate various stops and levels in the mine. Empire Mine. Collection of Ron Bonmarito.

Below:
The blacksmith shop of the Empire Mine, where everything was kept sharp and in good repair. Collection of Ron Bonmarito.

The mine head in the background with the mill. Empire Mine. Collection of Ron Bonmarito.

The superintendent of the mine lived pretty well. This was the boss's house when he was on site. Empire Mine. Collection of Ron Bonmarito.

A big mine needed good repair shops. This appears to be a carpentry shop. Empire Mine. Collection of Ron Bonmarito.

Interior view of the super's house, chiefly notable for the library and the African hunting trophies. Empire Mine. Collection of Ron Bonmarito.

NEVADA

Virtually at the back door of California's mother lode district, the Carson Valley area of Nevada served as a rest stop for increasing numbers of footsore travelers coming across the Great Basin to the fabled California gold fields. Just a few miles away—actually on the notorious Emigrant Trail to Placerville (then known as Hangtown, California)—was an area known today as Dayton, Nevada. Dayton sits at the terminus of a particularly bad stretch of desert to the east and the exit of the Carson River from its canyon. Dayton is also within fifteen miles of Gold Hill, Silver City, and Virginia City, famous names in mining.

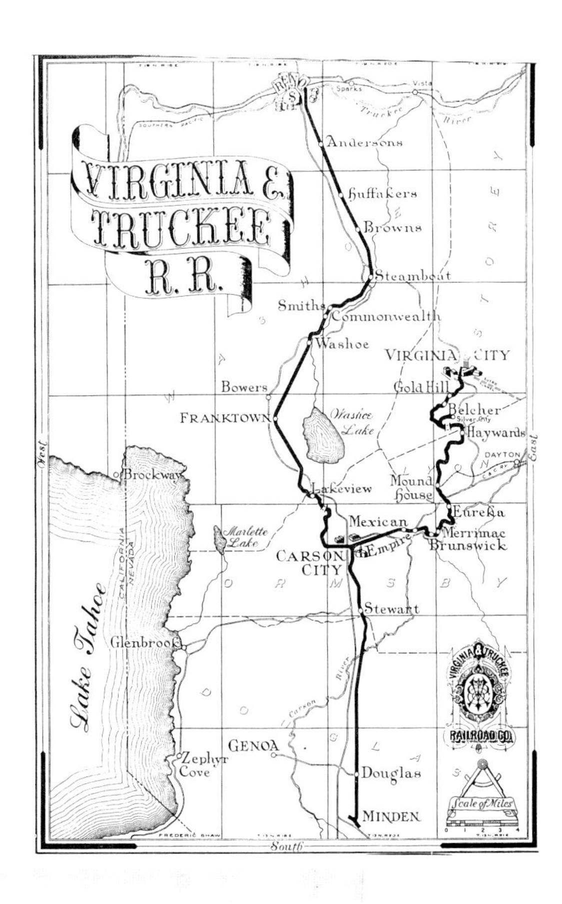

A period photograph of Virginia City, Nevada, taken from the far eastern end of the city. Mount Davidson is to the right with Six Mile Canyon to the far left and Gold Hill over the horizon in center. This photograph dates from about 1878 and clearly shows the two churches in the center. The Virginia and Truckee railroad tracks are just to the right of the church. Virginia City had a stable population of around 25,000 people at this time. Collection of Ron Bonmarito.

Originally established as a rest stop for the weary, Dayton and its outskirts were home to early Chinese residents who had taken to panning small amounts of gold from the nearby river. As early as 1851 (when the town of Genoa was founded as Mormon Station, the first town in Utah Territory, now Nevada), gold exploration was taking place throughout the general area. Although occasional small pockets were found, there was nothing significant enough to generate any kind of boom or rush. In early 1852, brothers Hosea and Allen Grosh began placer mining in Gold Canyon, which is adjacent to the mountain ridge where the Comstock Lode would be found. The Gorsh bothers took out small amounts of both gold and silver, but they both died of accidental causes in 1857 before developing their prospects. Others took over their claims but were bothered by the clogging of their riffles and quicksilver by an annoying blue clay—which, ironically, was eventually identified as sulpherites of silver.

In January of 1859 James Fenemore and Henry Comstock prospected the upper end of Gold Canyon and located many placer sites, still unaware that the richest ores were just below the surface. In early 1859 Peter O'Riley and Patrick McGlaughlin began placer mining at the head of Six Mile Canyon. While digging out a water reservoir, they discovered a lode that would eventually be known as the Ophir Bonanza. And finally, that summer, a chunk of the troublesome blue clay was taken to Grass Valley, California for an assay. There, it was calculated to contain over $3000 in silver and over $800 in gold per ton—rich indeed! Word spread quickly through the declining gold fields in California, and the rush to Nevada was on.

In general, the first people to show up at a gold rush in the West were the ones who made the most money, simply because they staked the ground first. Some of them actually worked the claims (you had to do some of that anyhow to establish a legal right to the area—to "prove up" the claim). In so doing they either enhanced the value of the claim by finding good paydirt, or degraded the value by finding nothing at all. Occasionally they simply failed to dig deep enough, and sold out too soon. Whether the claim was promising or not, it was usually sold to some moneyed latecomer, sometimes for a princely sum, sometimes cheaply for speculation.

MINING & THE LAW

Around this time it became very apparent that there were some fundamental problems with the laws regulating who worked what piles of rocks. In Virginia City (as in other mining boom towns) the law was whatever the local miners all agreed to; there was no other law on the books, and no enforcement authorities in any case.

Problems arose because it was quickly discovered that the hot pay streaks were all underground. A surface discovery was literally just the tip of the ore body, which could extend in large sections and in all directions underground. Since lode claims were filed for a specific piece of ground as surveyed on the surface, how could the rights to underground areas be controlled? Miners following ore bodies soon came into conflict over how far the claim boundaries extended, and in what directions. Did the boundaries simply go straight down at a 90 degree angle from the horizontal surface, disregarding the actual angle of the ore body, or did they follow the ore body at whatever angle it happened to take?

The mountains of litigation and body of law which developed around this issue eventually led the U.S. Congress to pass the Mining Law of 1872, which is still in force to this day, some 123 years later. The ensuing litigation made the attorneys involved some of the richest of the rich to be spawned by the Comstock. They commonly took shares in the mines as payment for services rendered which they then sold at a later date at inflated prices.

In 1995, it is now the law to file either a placer claim (basically surface rights only) or a lode claim (underground rights only)

THE COMSTOCK SILVER LODE

The Virginia City Comstock strike was unique in mining history, since for the first time ore was being removed from a mountain in large sections. Eventually, whole portions of the mountain underground were removed, made possible by the in-

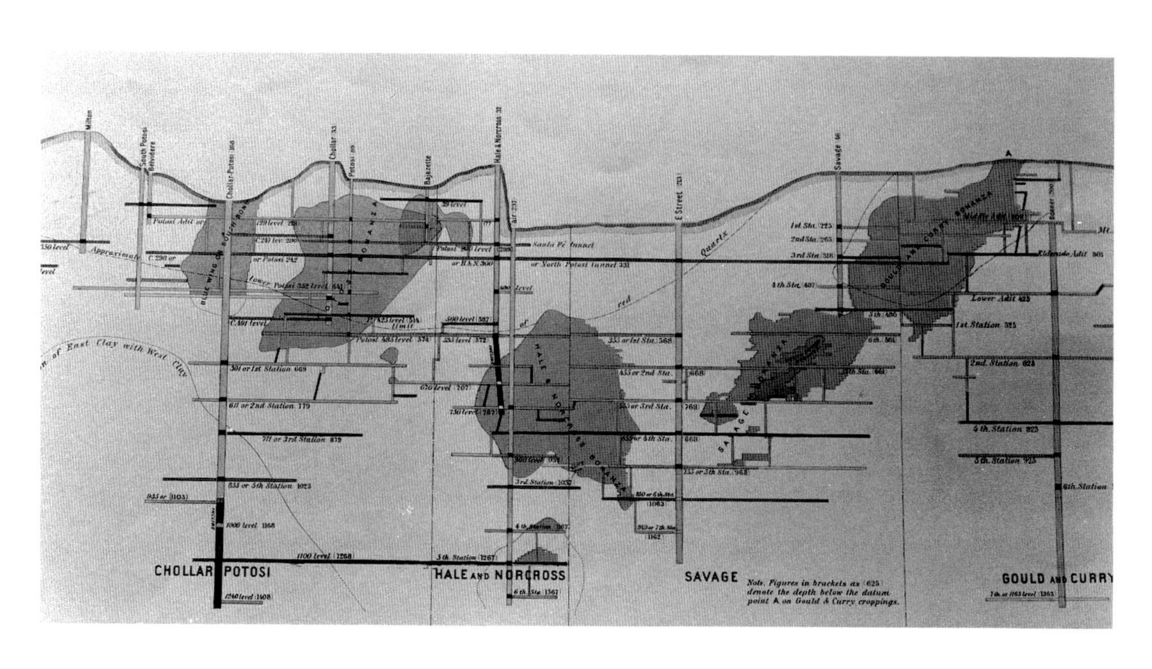

A longitudinal cross section of the Virginia City mines. Note the deepest shafts reach to about 1300 feet from the surface. The ore sections reach to the surface and span hundreds of feet. This map was done in the 1870s. Some of the mines eventually went much deeper, as far as 3000 feet. Collection of Ron Bonmarito.

vention of an enterprising young man named Philipp Deidesheimer. Deidesheimer developed a system called 'square set timbering', which proved so successful it was patented. It allowed the mining of dangerously massive, high-temperature ore bodies, which even in later years could not have been worked without it.

Deidesheimer's system involved creating a box frame of timbers without sides, tops, or bottoms, much like the erector sets of later days. In this way, timbers of managable size could be assembled to form a structure that was at once rigid, stiff, and very strong, but that still gave workers access to the ore bodies in all directions. The floors were planked over, and individual vertical sections were filled with waste rock to create support pillars. It was the perfect solution to the problem of keeping miners safe while they hollowed out large cavities (hundreds of feet long, wide and deep) inside a crumbling ore body.

This excellent photograph was taken sometime around the turn of the century in one of the Comstock lode's mines in Virginia City, Nevada. The photograph shows in detail the square set timbering system first used there and how it allowed the miners to work the ore body. Both men are using carbide lamps for light and ladders can be seen on the far right and lower left leading to other parts of the mine. Collection of Ron Bonmarito.

Photograph of a large poster originally printed to show off the square-set timbering system in use there. The square-set timbering allowed whole sections of the mountain to be replaced with props, and allowed the removal of large ore bodies that existed in the Comstock in no single vein to work. In this photograph the light-colored rock is the ore. Author's Collection.

In the Comstock mines, including Gold Hill and Silver City, additional problems eventually had to be surmounted. Not the least of these problems was water—lots and lots of hot water. Not only did the presence of subsurface water surprise the early miners (they were mining down from the sides of a mountain ridge in the *desert,* after all), but the temperature of the water was as high as 130 degrees Farenheit! This made the mines hellholes for the workers. Many of the the mining companies went to great lengths to eliminate the problems of excess water and heat. However, pumping water several hundred feet straight up required massive high-pressure pumps that were simply not available in this country. The initial answer to this problem was found in England, where monstrous 'Cornish' pumps had been used in coal mines for some time. These were purchased, transported around Cape Horn on ships, and dragged piecemeal over the Sierras by mule and oxen trains to Virginia City. Then large, substantial buildings had to be constructed to house the pumps, and steam power plants had to be built to power them. All of this did not come cheaply, but the fantastically rich silver ore was worth the expense.

Right:
This picture is labled as a mine station on the Comstock Lode, Virginia City, Nevada. In contrast to the lift station for ore carts in the Empire Mine in California, the Comstock miners filled a cart below the surface, moved the ore about in this cart while deep in the mine, and then lifted this same cart to the surface for dumping. This saved miners the work of emptying a sub-surface cart into a lifting cart. It was a lot easier on the carts as well. Also note the overhead timbering necessary to fill the void left by removal of massive ore bodies. Collection of Ron Bonmarito.

Left:
Possibly the earliest known photograph of a water-powered mill in Gold Hill or Gold Canyon. The water was piped from a reservoir higher up on the hill. This was not a practical solution because of the lack of water in the area. Collection of Ron Bonmarito.

This is for the Special Information of Gold Buyers, Officers, and Wells, Fargo & Co's employees only, and NOT FOR POSTING.

$1850 REWARD!

WELLS, FARGO & CO'S EXPRESS BOX,

On RENO Stage Co's Route, was ROBBED near VIRGINIA CITY, by 5 men, on the 3d inst., of Treasure as follows:

$1800 in Gold Dust, $600 in Coin.

In the Dust was a nice bright nugget, weighing 10 oz., 3 dwts. 1 1-2 inches in length, rough surface, but no quartz visible.

A REWARD of one-quarter of any Treasure recovered has, been offered, and will be paid for the same upon delivery to Wells, Fargo & Co., and

$250 EACH, ADDITIONAL,

For the ARREST and CONVICTION of the ROBBERS.

JNO. J. VALENTINE,
General Superintendent.

VIRGINIA CITY, Aug. 17, 1875.

Photograph of an original poster in the collection of Ron Bommarito. Interestingly, the poster refers to both gold dust and a nugget in 1875, when it would be assumed that the area would not have produced anything but refined gold by that late date. Collection of Ron Bonmarito.

LONGITUDINAL ELEVATION GOLD HILL MINES, COMSTOCK LODE.
SCALE: 200 FEET TO ONE INCH.

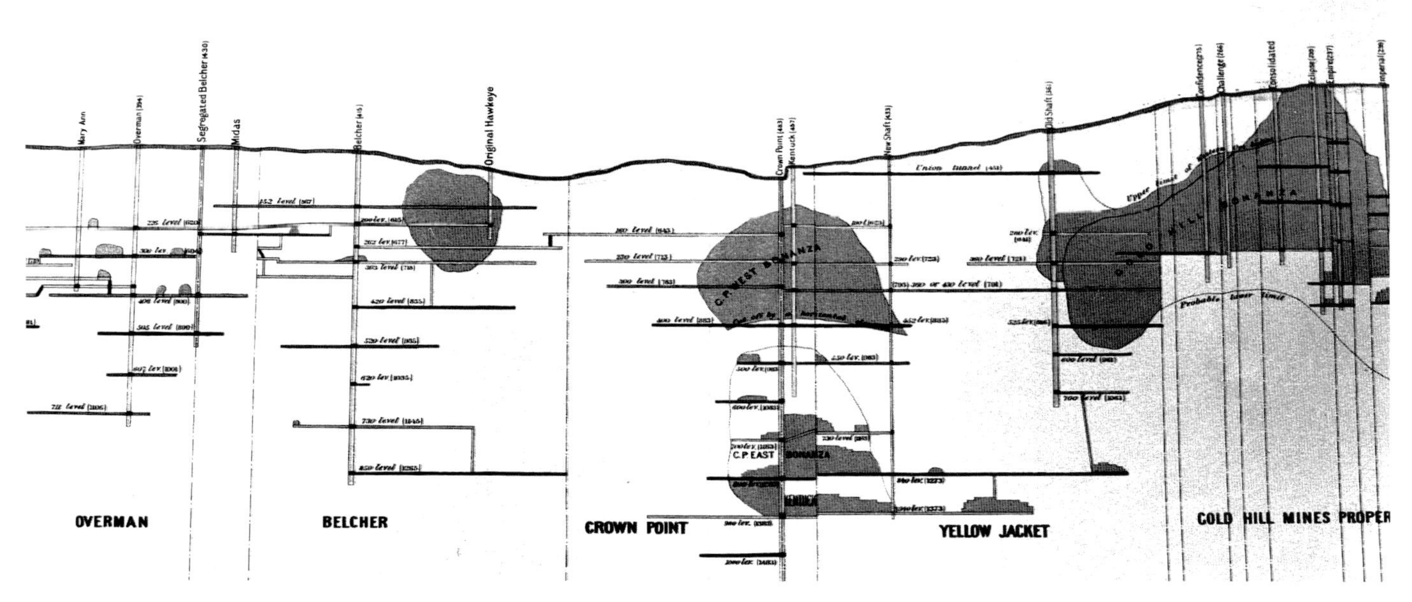

Gold Hill mines map, really part of the Comstock lode. It is easy to see that large ore bodies existed. Collection of Ron Bonmarito.

Dealing with the heat was another story! Being in a geologically young and active area has its downsides and one of these is hot rock (sometimes 150 degrees to the touch!). When the mine shafts were first extended downward it became apparent that human beings could not tolerate working in the temperatures that existed there. The mining companies brought in barrels of icewater, and required men to work as little as one hour at a time close to the 'face' (the working wall of ore). Then they were rotated upward to cooler zones, and then back down, until a full ten or twelve hour shift was accomplished. Because of these harsh conditions, the Comstock miners were well paid, as much as $4.25 per day.

These high wages were necessary; living in Virginia City was not cheap. As in any boom town, the cheapest commodities available were "booze and broads." Both were available in abundance in Virginia City.

The girls of "the line" were tolerated by all members of society but accepted by almost none. Prostitution was a necessary evil, invariably the first thing to be established in a boomtown. To their credit, the Virginia City girls seldom had the horrid reputations that other "soiled doves" did in the earlier days of the California gold rush. Indeed, many stories are told about Nevada "ladies of the night" who fell in love with miners, married, and had prosperous families despite their shadowy beginnings.

In Virginia City they had their own section of D Street, just one or two short blocks downhill from the main business and entertainment district. There were none of the stereotypical whorehouses of Western legend, as seen today in Hollywood portrayals of the Old West; Virginia City prostitutes were all individual operators in their own houses.

It is said that the second thing established in any boomtown (after the prostitutes) was a saloon. Whiskey and beer had to be available in abundance for the miners. Virginia City at one time had 150 saloons operating all at once. At its peak, it also had six police stations, four churches, and several music halls, as well as iron foundries and a couple of breweries. With a population that exceeded 25,000, Virginia City was a very busy place indeed.

Back in the mines, new problems developed, requiring new solutions. Ventilation shafts were dug or drilled to most of the mines to help circulate air into and out of the shafts and working areas. Eventually one enterprising engineer, Adolph Sutro, thought of digging a tunnel from the base of the mountain to the lower portions of the mines to drain the water out. As an additional benefit, this tunnel would allow miners to transport ore from the mine's lower portions straight out at the base, rather than having to lift it all the way to the top.

Two Virginia City miners eating dinner in their shack. Both men are sitting on buckets. Note the absence of frills. Except for the 1851 Colt and the cribbage board on the wall, life must have been pretty spare. Collection of Ron Bonmarito.

All of Virginia City's water was piped into the town from across the mountain, the water in the mines being unfit to drink. During the winter it occasionally got very cold in Virginia City: frozen pipes! Collection of Ron Bonmarito.

A rare and original photograph of the Millionaires Club in Virginia City, Nevada. There are twenty-six men in this photograph, which was taken sometime in the late 1860s. The club apparently had a floor of this building, which also housed a mercantile store and the Bank of California on the ground floor. The Bank of California was the single largest financier of the mines in Virginia City and was owned principally by a number of the mine owners. Collection of Ron Bonmarito.

This is the red-light row in Ely, Nevada, around 1920. The row had seen its better days by this time, and the women were servicing more railroaders than miners. Note the woman seated in the door at right and the ever-present stray dog in front of the bar. Collection of Ron Bonmarito.

While this was an admirable idea and was eventually implemented, the only person to make any money from it was Sutro himself. A superb salesman, Sutro convinced many investors to purchase stock in his tunnel-building company. While he had some money invested in it himself, he also drew a nice salary. Later, he quietly sold off his stock, at a huge profit. No one else was so fortunate. Building the tunnel took much longer to build than expected, and was much more expensive. By the time it was finished, the miners were already working below the levels where the tunnel connected, so it was never used for its primary purpose: draining the mines so the miners could work. The tunnel still exists, and still runs hot water out of the hills. Portions of it have collapsed, however, so it is very dangerous.

Interior of Sutro shaft hoist #2. The Sutro tunnel had several ventilation shafts during its construction. Collection of Ron Bonmarito.

Exterior of shaft house #2. Note the pile of tailings in the foreground. Collection of Ron Bonmarito.

Adolph Sutro's mansion. Built by Sutro during the construction of the tunnel, it is in close proximity to the tunnel and was to have been the focus of a town with the same name. Collection of Ron Bonmarito.

View of the Sutro mansion and its proximity to the machine shop. The tunnel is just to left out of the picture. Collection of Ron Bonmarito.

Interior view of the Sutro machine shop. The men were apparently showing off a newly built wheeled carriage for a compressed air drill, a very fancy piece of equipment by the looks of it. Note the Republican Party banners on the walls in rear. Collection of Ron Bonmarito.

Interior view of one of the 'Cornish' pump houses used in Virgina City, Nevada. The wheel is about 30 feet in diameter. They were capable of pumping water over a thousand feet straight up. Collection of Ron Bonmarito.

Interior view of the Brunswick Mill in Carson River Canyon about 1876. This and other mills in the canyon were connected to Virginia City by the V & T. Watkins photo. Collection of Ron Bonmarito.

A portal of the Sutro Tunnel, probably before it was finished. The gentleman in the center with the white beard is probably Adolph Sutro. Collection of Ron Bonmarito.

Virginia City, Nevada from the southeast, looking toward Mount Davidson. The large building in the foreground is the California Pan Mill, which was one of the larger mills in the city. This photograph dates from 1878. The slight blurring in the photograph is caused by damage to the original negative. Collection of Ron Bonmarito.

Like so many other Western mining towns, Virginia City is fairly high up in altitude, at about 6200 feet. As is the case with many mining areas of the Great Basin, it has cold, hard winters. There was no coal available, so keeping a boom city of 25,000 people warm was a tough job. Wood, in particular firewood, was always in demand, and in an area where there are very few trees this created a problem. What wood was available had to be moved fifty or more miles by mules or oxen. In 1866 it was estimated that some 200,000 cords of wood were delivered to the end consumers in the district[2]—a cord being a stack of wood measuring 8' x 4' x 4'. The town's estimated wood usage that winter was 568 cords per day! At roughly $10.00 per cord this worked out to about $2,000,000 per year.

This amount did not include the lumber and timbers used by the town for building, or in the mines for shoring, ties, etc. The town's domestic requirements at this time came to about 7 million board feet per year, while the mines themselves used in the neighborhood of 18 million, for a total of about 25 million board feet.[3] While this number would decrease over the years, the amounts used were still staggeringly high.

In addition to the town's fundamental need of firewood for keeping warm, and the general need for building timber, all of the mines ran their equipment on steam power generated by wood-fired boilers. Initially, a thriving population of brush gatherers made a living harvesting the abundant growth in the surrounding hills. This brush provided good fuel for the large boilers then in use, because it burned with a very hot flame and was easy to get. However, it soon required longer and longer trips to find suitable brush, and whole sections of ground were virtually denuded. Wood and brush were in short supply, and teamsters and wood gatherers were making a good living!

Most or all of the wood imported into Virginia City came by freight wagon at first, largely from the nearby Sierras. Much of the logging took place around the Lake Of The Sky, Lake Tahoe. The timber there was large, old-growth wood that was brought to a central point and moved over the mountains by a water flume, then hauled overland to Virginia City. Additional supplies of firewood were floated down the Carson River from the California Sierras, then hauled up the mountain on large wagons built for this purpose.

Keeping a steady supply of firewood was a continual problem. It would appear that these young entreprenuers found a way to pick up scrap wood and haul it to Virginia City for sale. Collection of Ron Bonmarito.

This was what it was all about. Shown here is one week's production from a single mill, The Mexican, in Virginia City. It isn't known if this was gold or silver or a combination, but either way it is a lot of bullion. Collection of Ron Bonmarito.

MINING AND THE RAILROADS

This whole supply system was greatly enhanced in 1869 with the building of the Virginia and Truckee Railroad. The V & T was an instant success because it could carry wood up the hill to Virginia City, and ore back down, transporting it to the mills along the Carson River. This one railroad allowed the mines to work with the lower-grade ores that had been discarded at first by the mills in Virginia City as too costly to process. For the most part, these ores were processed by mills located some sixteen miles away in the Carson River Canyon. The location was significant then, as it is now—for today the river remains so polluted with mercury that fish taken from it are completely unfit for human consumption. Mercury was used in the amalgamation of the ores as it combines readily with gold and silver and can be driven off and recovered by a simple retort smelting process, much like the distillation of water.

Gold Hill, Nevada, looking toward the ridge that divides it from Virginia City. The tracks in the foreground are the Virginia & Truckee main line, which wound around the hills above and into Virginia City. The mines here were very rich, and were concentrated into a relatively small area. There is no mining going on here at the moment and most of the buildings are gone, as is the V & T track and tressle. Note the large stacks of firewood. Collection of Ron Bonmarito.

VIRGINIA & TRUCKEE R. R.

ON AND AFTER SATURDAY, FEBRUARY 1, 1873,

THROUGH PASSENGER TRAINS

For RENO, connecting with Eastern and Western bound Passenger Trains of the C. P. R. R., will leave

Virginia City Station.....at 9.00 P. M.	Reno..................at 2.25 A. M.
Gold Hill..............at 9.12 P. M.	Carson City........ ....at 4.05 A. M.
Carson City.............at 10.45 P. M.	Gold Hill..............at 5.20 A. M.
Arriving at Reno.........at 12.15 A. M.	Arriving at Virginia City..at 5.30 A. M.

LOCAL PASSENGER TRAINS

LEAVE

Virginia City for Carson City,	Carson City for Virginia City,
At 7.45 A.M., 11.45 A.M., 8.45 P.M.	At 8.00 A.M., 12.00 M., 4.00 P.M.

FREIGHT AND ACCOMMODATION TRAINS

Leave VIRGINIA CITY for RENO at 7.45 A. M. and 3.45 P. M., and will leave RENO for VIRGINIA CITY at 8.00 A. M., [illegible] P. M., and 5.30 P. M.

FREIGHT TRAINS

Leave VIRGINIA CITY Daily, at 5.45 A. M., 7.45 A. M., 9.45 A. M., 11.45 A. M., 1.45 P. M., 3.45 P. M., and 5.45 P. M.

Leave CARSON CITY Daily, at 6.00 A. M., 8.00 A. M., 10.00 A. M., 12.00 M., 2.00 P. M., 4.00 P. M., and 6.00 P. M.

At Virginia, Gold Hill and Carson Offices, Through Tickets can be obtained to SAN FRANCISCO, OGDEN, and intermediate points on the C. P. R. R.

Through Tickets can be procured at the Company's Office, Virginia, for CHICAGO, KANSAS CITY and ST. LOUIS.

H. M. YERINGTON, Gen'l Sup't.

This is from an original Virginia & Truckee Railroad advertisement. The schedule is bewildering, but it clearly shows that a lot of trains ran from Virginia City to Carson City and points beyond. Collection of Ron Bonmarito.

The original south end of the V & T in Carson at the Carson & Tahoe Lumber & Flume Co. yard. The yard was just south of Carson City and the terminus of the flume from Lake Tahoe where all the timber for the Comstock was cut. The flumes to left and right brought the lumber in where it was stacked, then shipped by the V & T to Virginia City. Collection of Ron Bonmarito.

The Virginia & Truckee main station in Carson City about 1890. The V & T was headquartered in Carson with its shops and foundry two blocks away from this depot. This building is still there but all other vestiges have been torn down. Collection of Ron Bonmarito.

The Virginia & Truckee met up with the narrow gauge Carson and Colorado RR in Mound House, a few miles east of Carson. The Carson and Colorado went south, going all the way to Bishop, California and then finally to Keeler, California. Collection of Ron Bonmarito.

A real Concord coach still in use but in rough shape, probably heading east from Mound House, Nevada. Dated somewhere around 1890. Collection of Ron Bonmarito.

Photograph of the poster showing the various protagonists in the only known train robbery of the Central Pacific Railroad in Nevada. The men were caught. Collection of Ron Bonmarito.

Fabulous fortunes were made and lost as the result of the Comstock lode of Virginia City. Most of the money went to San Francisco, where *all* the real money had gone since the beginning of the Gold Rush; ownership and financing for Western mining was seated there. By the time the Comstock was reaching its peak, the California mines were slowing down; but for San Francisco, what the gold rush of '49 had started in California was brought to a head by the silver fortunes of the Comstock. Moreover, the trip between Virginia City and San Francisco was an easy train ride. The constant coming and going of the trains running full both ways proved to be a bonanza for the mines, for the mills, and for the Virginia and Truckee. Everybody made a profit.

Ultimately, what the world began to realize was that there were huge mineral resources in the Great Basin and the surrounding mountains and they were just waiting for the lucky persons to find them. With this realization early on, the prospectors and speculators fanned out in all directions, for gold (and silver) is where you find it.

OTHER NEVADA MINES

Exploration continued at a steady pace elsewhere in the state of Nevada, and many discoveries were made. Usually, when a significant discovery was made a tent camp would spring up. Since there was little or no timber to be had in these essentially desert locales, the town's permanence depended on the richness of the strike. The better the strike, the better the quality of the town that grew up around it. It was simple economics, really; no sane businessman would invest time or money in putting up a permanent structure if it looked like the prospect was not going to last more than a year or two.

By 1862, what would come to be known as the town of Austin, Nevada was one of these places. Located about 180 miles east of Reno (about 160 miles east of Virginia City), it was quite literally in the middle of nowhere. The growing town faced many logistical problems, since everything needed to start mining and building had to be transported a very long distance by horse-drawn freight wagons. To justify such an effort, the ore needed to be very rich; and in the case of the town of Austin, it was. Some assays calculated that the ore contained as much as $6000 per ton in silver alone. In 1864 Austin was incorporated, and very soon it had all the trappings of a modern city.

What happened to the Austin area next is typical of this kind of discovery; prospectors fanned out in all directions, and made many other smaller strikes. Each time ore was found, the same scenario was enacted; a camp was established and the appearances of civilization were started. Often, the individual mines were depleted (they were "played out") and the miners moved, taking their town with them—literally. Small wooden cabins could commonly be seen moving through the countryside on freight wagons, either intact or in parts.

An early photograph of Austin, Nevada, date unknown, but probably in the 1870s. This photograph was taken from the mountain ridge behind the town facing east. Most of the mines are actually beneath the town, with tunnels branching into the mountain itself. Except for the mills, which are all gone, the view today is not much different. Collection of Ron Bonmarito.

Austin, Nevada. The Old Manhattan Mill produced over $19,000,000 during its lifespan, almost all in silver. Collection of Ron Bonmarito.

The Austin Tunnel. This tunnel is in the face of Lander Hill, which goes directly under the town of Austin some 6000 ft to the Frost Shaft. This was the center of the mines on Lander Hill and it had shafts that ran 3500 ft left and 2800 ft right to connect with other mines. Collection of Ron Bonmarito.

Austin, Nevada. The Old Boston Mill preceded the Manhattan Mill and was destroyed by fire. Collection of Ron Bonmarito.

This is the compressor building for the Austin Mines. Take note of the volume of water flowing out of the mine in foreground. Unlike Virginia City, it was possible to drain the mines in this manner rather than pump them out from above. Collection of Ron Bonmarito.

A shift going into the Austin Mine in ore cars. Notice the variation in ore cars in use. This was fairly typical of mines in this area. Collection of Ron Bonmarito.

Austin Mine. This is described as an area near the mouth of the tunnel where 'bad rock' required extra timbering. Note the horse's head in center at rear. The men in those ore carts had best kneel down in this section. Collection of Ron Bonmarito.

EUREKA

Directly east of Austin is Eureka, Nevada. Like Austin, Eureka had ore bodies of a very high grade, and like Austin was situated in a mountainous ridge that led to some cold winters and rough living conditions. The area between Austin and Eureka consists of high plains, broken by mountain ridges running mostly north and south. Almost all of the resulting valleys are above 5000 feet in elevation, and can boast of little more than scrub brush, sage brush, sand, and rattlesnakes. The area is very hot in the summer and frequently very cold in the winter. Eureka is much like Austin in another respect: it is a long way from anywhere, and difficult to live in even if you were involved in high-grade mining.

In some respects, Eureka is very different from most other boomtowns. Eureka's citizens were quick to develop an opera house and other substantial brick structures, many of which are still standing, some in completely restored condition. Among these is the county courthouse, which is still in use. The town erected permanent structures like these because bricks could be produced locally with the area's supply of clay and firewood. Had the townspeople had to transport the heavy bricks from a distance, it is not likely that their early buildings would have been any different from the ramshackle wooden structures set up in most new boomtowns.

Eureka remains a viable city today because of mining revenues still being generated by new mines, which utilize new techniques and low-grade ores passed over in years past.

Eureka, Nevada, 1912. The day shift poses for the camera. Virtually all of these men are known, and most are Italian. The black man in rear, second from right, is unusual in the western mines. Many of the descendents of these miners are still in Eureka today. Collection of Ron Bonmarito.

Eureka, Nevada on July 4th, 1897. The whole town turns out for the big parade. This photograph was taken from the balcony of the opera house looking east. The main street today is the paved U.S. Hwy 50, but some of the same buildings are still there, including the opera house which has been restored. Collection of Ron Bonmarito.

An early photograph of Eureka, Nevada looking south west. The major mines were in the hills to the right. Note the little boy at the left of picture. Collection of Ron Bonmarito.

A later photograph of Eureka from about the same angle as in the previous picture. Buildings are now brick. It gets COLD in Eureka in the winter! Collection of Ron Bonmarito.

Despite the difficult terrain and challenging climate, Nevada's early explorers and prospectors wandered north and south from the middle of the state in search of more gold and silver. They kept finding it, mostly in small strikes and mostly in very out-of-the-way places. They continued due east from Eureka into what would become White Pine County and into the area around Ely (or what would become Ely). Some of the towns sprang up because of the discovery of very rich ore that was almost pure silver. Near what was to become the town of Hamilton, a mine known as the Consolidated Eberhart Mine took ore worth $3,200,000 from a service pit 70' x 40' x 28' deep. The ore assayed out to over $1,000 per ton. Another boom was on!

The Hamilton area is high up in the mountains, nearly 8000 feet in elevation. The town grew so fast that miners were hard pressed to find places to sleep other than on the ground. One story tells of two miners who hastily built a rock cabin for shelter one winter, only to find in the spring that the rock they had used was high-grade silver ore. They sold it for $75,000![4] The ore in Hamilton was found in large chunks, including the largest boulder of 'horn silver' believed ever to have been found. Horn silver is very high grade, dull gray ore, which can be seen by the naked eye to be something out of the ordinary. In some cases, it looks like extruded wire. The gigantic orn silver boulder found in Hamilton was said to have weighed in at about 40 tons.

By 1868 some 15,000 people had boomed their way to Hamilton to work the mines. The growing town needed lumber for building and firewood for burning, and the mines of course needed timbers and steam furnaces themselves. One enterprizing teamster hauled a large load of lumber to Hamilton from far off Genoa which is on the western edge of the State. He was paid the high price of 12 1/2 cents per pound for freighting the wood the distance of about 200 miles.

In most mining towns, the only available women were the prostitutes; Hamilton was not much different. The only female company for the Hamilton miners were two Mormon girls, whose mother refused to let them go to a planned dance—primarily because they had no shoes, and the dance floor was rough sawn lumber. The problem was quickly solved when somebody donated two pairs of men's shoes that almost fit the girls. Still, the dance lasted only so long as the fiddler stayed reasonably sober, which in this case was about an hour. Then he fell off his stool, out the back of the tent-style building, and down some six feet to the rocks below, where he broke his neck. Then the dance ended.

Hamilton, like so many other boom towns of the West, lasted only a few years. By 1876 the Eberhart mine was completely abandoned. It was reported that as late as 1897, there were four boarding houses in town, all owned by one woman—but she only had two renters!

Hamilton was west of Ely; Ward, on the other hand, was a few miles east. Ward had the distinction of being one of the dirtiest mining towns in existence, partially because the town was at a high elevation (it was above 8000 feet, where rain seldom fell to wash everything down) and partially because of the number of charcoal ovens. Charcoal ovens had been established in Ward to take advantage of the local trees for the production of coke (really charcoal). Coke was used in for smelting gold and silver ores and for reducing amalgam (mercury with gold and silver) into its separate parts. By 1875 Ward had a population of about 1500 people, and grew to 2000 people at its peak.

Ward must have been one of the wilder towns in this wildest of areas. One old-timer is quoted as saying, "There were so many burros on the streets that one had to climb from one to the other to get across to the opposite side."[5] The rougher element of society was there in abundance; this included 'ladies' known as Shoofly Minnie and Big Mouth Anna, who were housed in a place called Frog Town. No explanation for the name of Frog Town is offered.[6]

The mines here had to have been some of the worst anywhere. One account reports that "Men were overcome by the gases after a round of dynamite shots. The miners, returning to work too soon after a round of shots had been fired, would be overcome before they could reach better air. They would ring an alarm signal which would start an outside crew...into a tunnel. The asphyxiated miners would be loaded onto a hand car like so much cord wood, taken to fresh air, and revived."[7]

Because the ores also contained arsenic, many of the miners and populace were slowly poisoned by the smelter fumes containing a witch's brew of arsenic, mercury and lead fumes. It was said that vegetation refused to grow in the area for many years after the mines shut down.

SEVEN TROUGHS

Independence Day was a big deal even in the smallest of towns. This parade is in the town of Seven Troughs, Nevada. The year is 1908 or 1909 and there appear to be more people parading than watching. Collection of Ron Bonmarito.

Seven Troughs, Nevada, on the 4th of July. The band has come back and is now playing on the porch of the saloon. Note the child on the porch in a sailor suit and what appears to be an Indian at far left. Collection of Ron Bonmarito.

Everybody dresses up for the 4th of July. The restaurant is across the street from the saloon in Seven Troughs. Collection of Ron Bonmarito.

The Seven Troughs baseball team. Small towns took their baseball seriously and everybody came to see the team play. Collection of Ron Bonmarito.

Seven Troughs. After the parade and the baseball game people got together for parties. The accomodations were typical for these desert mining towns: rough. Collection of Ron Bonmarito.

The man at far left appears to be drunk, if the disapproving looks of the women are any indication. Seven Troughs, Nevada. Collection of Ron Bonmarito.

The town of Seven Troughs, Nevada. The town was in existence from 1907 to 1918. No indication of where the baseball diamond was located. Collection of Ron Bonmarito.

One of the hazards of life in the desert is the flash flood. A few women appear to be picking through the debris after a major portion of Seven Troughs was damaged or destroyed in a flash flood in 1915. The mines played out soon after and the town was abandoned by 1919. Collection of Ron Bonmarito.

RAWHIDE

One of the classic boom towns of all time, Rawhide, Nevada. The boom lasted for just two years although the town continued to exist until the 1940s. It was started in 1906 when this photograph was apparently taken, and was finished by September 1908 when a fire destroyed most of the town. Collection of Ron Bonmarito.

Main Street in Rawhide, Nevada. The wagon in left front appears to have high-grade ore sacked for delivery to the nearest railhead. The wagons in the background seem to have trade goods coming into the town. Collection of Ron Bonmarito.

Loading high-grade ore from the Kearns Lease in Rawhide, 1908. Leasing claims was a common way to work a mine. The original filers of the claim would lease out a claim for a percentage of the take while they were out looking for more claims. Collection of Ron Bonmarito.

Rawhide, 1908. "Pals Asleep." From an early postcard mailed in Rawhide. Collection of Ron Bonmarito.

The arrival of a stage in Rawhide was always a big event. In this case the stage is almost swamped with riders. Collection of Ron Bonmarito.

Rawhide burns! September 4th, 1908 and most of the town was destroyed by fire, always a threat in water-parched interior Nevada. Collection of Ron Bonmarito.

THE BIG BONANZA

Meanwhile, the mines back in Virginia City—which had begun to decline after ten years of good production—were suddenly invigorated by the biggest discovery of all, appropriately called The Big Bonanza. The Consolidated Virginia mine in 1873 brought in a very deep lode of high-grade silver ore, and the VC mines once again had a new life. This one mine alone paid out over $75 million in dividends to its owners through the years. It was the largest ever of the Virginia City mines; several smaller mines which had been working the same lode from different shafts had settled their legal problems by consolidating, hence the name *Consolidated* Virginia.

MORE MINING DEVELOPMENTS

To illustrate the scale of production at Virginia City and throughout Nevada, we can quote contemporary sources. The Centennial Gazeteer of the United States gives us the following numbers: of an aggregate production of gold and silver in the entire United States of $1.21 billion during the years 1848 to 1868, California provided $900 million, Nevada $90 million, with the remainder divided between Montana, Idaho, Washington, Oregon, Colorado, New Mexico, and Arizona. For the year ending December 31, 1867 the totals were more indicative of what was actually happening; California had slipped to $25 million, Nevada was up to $20 million, Montana provided $12 million, and the other states contributed a combined total of $17 million. By 1872 the numbers began to change dramatically. Nevada produced $25.5 million, California had dropped to $19 million, Montana had slipped to $4.4 million, while Utah showed up with $3.5 million. During the same year Arizona slipped to $143,000, and the other states had begun to fall rapidly.[8]

THE BODIE MINES

Bodie, California is a mining town just over the California/Nevada border. Mining journals generally refer to Bodie in their discussions of Nevada, because access to the town in the days of discovery was almost always from the north and east—in other words, from Nevada. Initial prospecting work was done as early as 1859, when gold was discovered there. Isolated as it is in the eastern mountains of California, Bodie is a distinctly separate mineral source and not at all part of the famed Mother Lode. Because it was found at an elevation of nearly 9000 feet in a high alpine setting, it was a long time before serious mining took place there. It didn't help that a number of people had died of exposure there in the earliest years!

The mines at Bodie produced some gold and silver ores of very high grade, but not until 1876 and 1877 did serious production commence, largely because of the town's remote location. Bodie only lasted about three years as a major producer, and then went into a rapid decline. Most or all of the "boomers" left when the big bucks quit coming out of the ground. Still, the town managed to hang on until the 1930s, when it was abandoned lock, stock, and barrel. Today it is one of the truly pristine ghost towns available for tourists to view (but only in the summer; winter completely closes all access to Bodie). The town is protected by California State and Federal Rangers for the world to see.

The Wells Fargo & Co freight office was always in evidence even in the smallest towns. This one was in Rawhide, Nevada. Collection of Ron Bonmarito.

THE AURORA MINES

Across the border from Bodie lies the town of Aurora—or, more accurately, the location of the town of Aurora. Established in 1862, Aurora was originally thought to be inside the state of California. The earliest mines were in fact taxed by that state. It was not until late 1863 that an ongoing border survey proved that Aurora was in Nevada. Finally, the citizens of the town could dispense with one of their two elected sheriffs; before then, they had maintained one sheriff for Nevada and another for California.

The town of Aurora was built of substantial materials, for the inhabitants were aware of the risk posed by fire. The bricks were brought in by freight wagon, and were the reason the town lasted as long as it did. But alas, the deposits (though rich) were all on the surface. After producing some $30 million in about ten years, they were completely played out. During the 1890s the post office was closed, and the town slowly emptied out. A couple of revivals of the mines were attempted later, in the twentieth century, but nothing much came of them. By the late 1940s, the town was completely gone. Even the bricks had been removed; the entire town had been privately owned, and the owners sold off the used brick as an attractive building material. Today, even the town cemetery has disappeared from view.

Aurora, Nevada, sometime after the turn of the century when the town was in decline. Collection of Ron Bonmarito.

Aurora, Nevada and another 4th of July celebration. This was taken post-1905 judging by the early auto in right center. Note that the trees in front of the stores have been cut and tied to the porch uprights for decoration. Collection of Ron Bonmarito.

THE TONOPAH MINES

Not far from Bodie as the crow flies (about 100 miles) is a rough desert mountain that seems to rise out of the surrounding sand and salt flats like a castle on a primitive island. It was here in 1900 that Jim Butler, a previously luckless prospector and a noted tall-tale teller, happened to find one of the richest strikes in Nevada. Legend has it that one morning after making camp on the mountain, "Big Jim" (as he came to be known) was searching for his wandering burro. In the process of chasing him down he noticed that the animal had kicked over a rock that appeared to be very high-grade silver ore. When he took the time to look a little harder, he discovered that the ore was very rich indeed. Eventually he filed claim to most of what came to be known as Tonopah, Nevada. He, like so many others, sold out for cash before the really big dollars were made, but not before the legend of his burro's lucky kick was started.

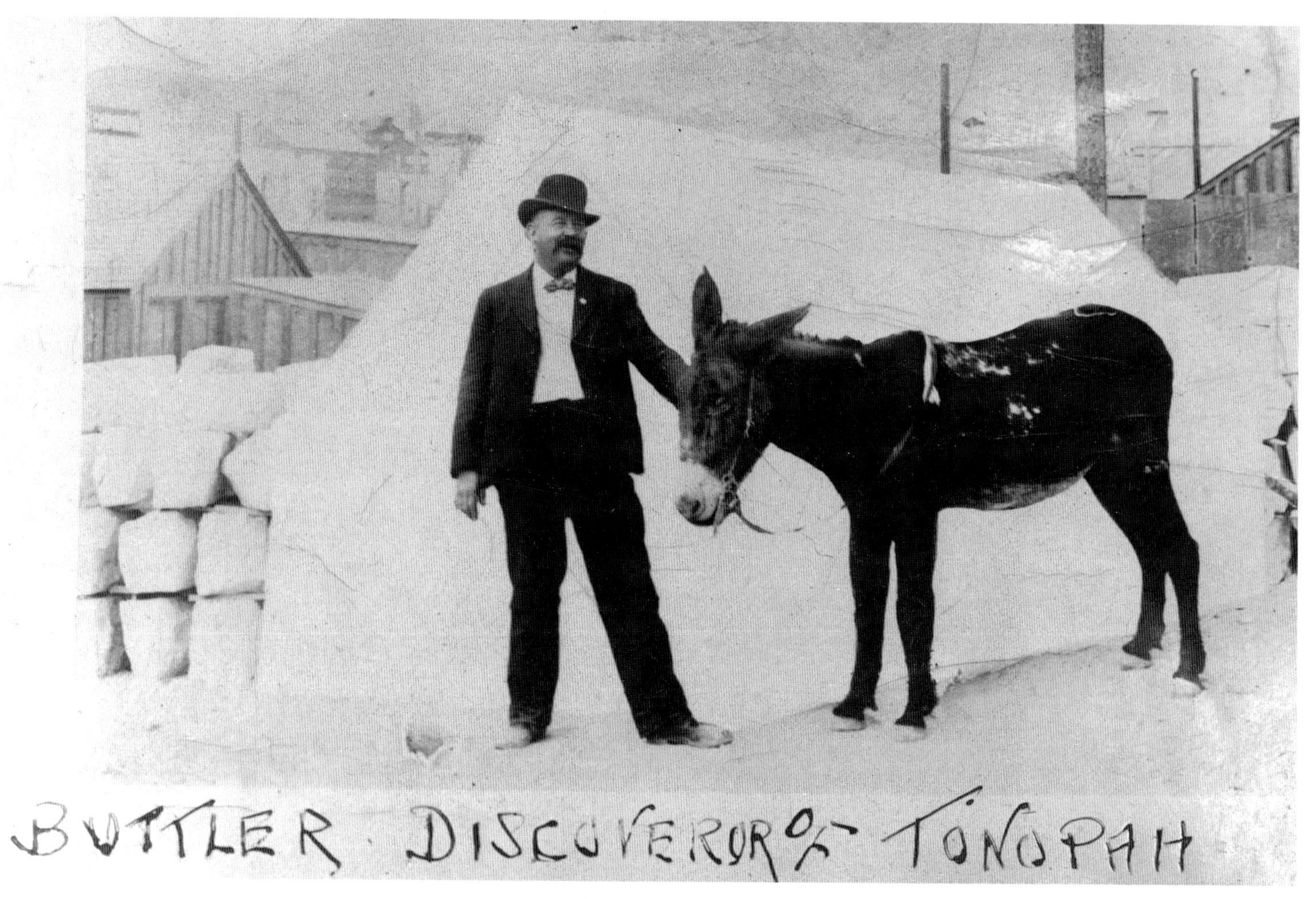

Jim Butler and his mule. The discoverer of the Tonopah, Nevada lode. Butler's name is misspelled on the bottom but spelled correctly elsewhere on a sign. Collection of Ron Bonmarito.

Mining began in earnest almost at once and in 1902 the Tonopah Mining Co. began both surface and shaft mining. The ore at Tonopah had to be treated chemically to release its treasure, requiring massive mills for which investors were only too happy to provide financing. The mines here continued to very productive into the 1920s, and did not stop altogether until the Tonopah Mining Co. stopped working in 1948. The largest producing mines were the Tonopah Mining Co. at $48 million, the Tonopah Belmont at $38 million and the Tonopah Extension at $22 million.[9] These three mines accounted for two-thirds of the entire production of the area, which had a total of sixteen mines. Gross yield was something over $150 million during those years. The town of Tonopah exists today as a stopping point on the main highway from the northwest of Nevada to Las Vegas and as a supply point for ranches and mines to the east.

Down the road from Tonopah was the next discovery of note, at Goldfield, Nevada. Today little more than a slow spot in the road, Goldfield during the early 1900s was the equal to Tonopah. The town's mines produced high-grade gold ore from the surrounding hills.

What made both Goldfield and Tonopah viable was the railroad which was pushed south from Hawthorne, Nevada south across the desert to Tonopah, then an additional 30 or so miles to Goldfield. This allowed ore to be processed in existing mills in other parts of the country, and made it feasible to bring in large equipment for use in the mines at reasonable rates.

Four telephone linemen. This photograph is dated 1906. The young man on the far left has what appears to be a Colt Bisley revolver stuck in his belt. Collection of Ron Bonmarito.

A genuine twenty-mule team freighting into Tonopah. All early freight came into Tonopah this way. Collection of Ron Bonmarito.

A twelve-horse hitch pulling freight into Tonopah. Collection of Ron Bonmarito.

The express wagon which ran between Tonopah and Goldfield before the days of automobiles and trains. Collection of Ron Bonmarito.

Goldfield, like virtually all other mining boom towns, began as a wide open place. It became the location for a number of prize fights promoted by Tex Rickard and others. Why anyone would travel all the way out in the desert to watch one championship fight is beyond this writer. The promoter was obviously very adept at his craft, however; there are many photos showing rather good crowds in Goldfield on hand for big-dollar fights between big-name fighters. Gambling—neither legal nor illegal at this time—was of course also rampant.

Although it appears barren to the untrained eye, this entire desert area was (and still is) a storehouse of minerals and chemicals which could be mined profitably. An enormous number of small towns sprang up and disappeared as ore strikes were made and mined out. Most all of these have disappeared with little trace other than a note in a book and perhaps a few photographs. In some cases the name alone exists, with no photos to substantiate the record. If the town existed long enough for a mill to be built, the concrete foundations are frequently all that remains. If intrepid travelers have the time and equipment to search, they can follow railroad grades. However, loose sand, snakes, and old mine shafts pose dangers to explorers, as do dehydration in the summertime and hypothermia in the winter. This is a perilous sport, and is not recommended for novice desert wanderers.

One of the more remote towns that existed to the east of Tonopah was Tybo. Located in a canyon above an area of wide open desert, Tybo was born in 1874 after an 1870 silver find. The ore here was an abundance of lead-silver, but it was a long way from any market. Most of Tybo's production went to the West Coast, and the town continued to be a serious producer through the 1880s. It was basically a quiet camp, except for the one time when someone imported Chinese laborers to cut wood; the locals banded together and chased them out of town! Tybo hung on until the 1930s, but is now almost a ghost town, with only one occupant and a few seasonal inhabitants.

No indication of what the occasion was for the photograph, but the sign in the background is for a brewery, which would be one of the first for Tonopah. Collection of Ron Bonmarito.

No indication as to whether this burro pack train is coming or going, but the 5-gallon cans on their backs appear to be full of whiskey. Collection of Ron Bonmarito.

The 4th of July in Tonopah was a time for celebration, and rock drilling contests were a big draw. A timed event, these two men are double-jack drilling into a block of granite in the platform. These events were big affairs with major prize money offered. Contestants from all over the West entered. The grandstand in left rear is for ladies only! Collection of Ron Bonmarito.

The year is 1911 and the place is Tonopah. The Valley View Mine has adopted the square set timbering system to work the ore in this mine. This is a good photograph of how the timbering system looked and how the miners worked around it. They appear to be chipping loose ore from the roof of the workings. Collection of Ron Bonmarito.

By 1913 the Belmont Mine and Mill was the biggest in Tonopah. This mill was just around the side of the mountain from the earlier photograph views and continued to operate until the 1940s. Collection of Ron Bonmarito.

One of the oil-fired locomotives of the Tonopah and Goldfield Railroad that connected both cities with the outside world to the north. The lady in the cab was not the regular engineer. Collection of Ron Bonmarito.

An apparently new mine head in Goldfield, Nevada. We don't know what the occasion is but the three men on the right seem to be passing around a bottle outside the privy, the tin shack to the right. Collection of Ron Bonmarito.

Mary's Dance Hall in Goldfield was a "house of ill-repute" around 1910. The cribs or ladies' rooms were to the left of the main doors. Collection of Ron Bonmarito.

There is no date on this photograph but it appears to be of Goldfield around 1910, with a number of substantial brick buildings under construction. A Larson photograph. Collection of Ron Bonmarito.

This photograph is dated 1904, shortly after the founding of Goldfield. The numbers indicate the locations of some of the major mines. Contrast this photograph with the previous one. Collection of Ron Bonmarito.

The Redtop Mine in Goldfield. An early photograph, indicated by the practice of sacking high-grade ore to be taken to a mill, possibly as far away as Carson City. This picture was taken by a photographer from Pomona, California named Larson. Larson took many pictures of the Tonopah/ Goldfield area. Collection of Ron Bonmarito.

The Combination Mine in Goldfield. The railroad in the foreground is probably the Tonopah and Goldfield, although others eventually served the area. All are now gone and forgotten. Larson photograph. Collection of Ron Bonmarito.

Inside the Mohawk Mine in Goldfield, Nevada. Collection of Ron Bonmarito.

Two intrepid prospectors loading their burros. These long-suffering animals were unaffectionately known as Rocky Mountain Canaries, for their unmusical braying. Collection of Ron Bonmarito.

Date and location unknown, but probably outside of Tonopah around 1905. The chief item of interest here seems to be the two all-white burros in the center of the photo. Note also the amount of trash and tin cans lying about on the ground. This is what was left after a boom town collapsed. Collection of Ron Bonmarito.

Passing freight wagons on the desert. It was unusual to see a wagon with the round 'prairie schooner'-type cover in this photo. Note that there also seem to be several wagons in the line in background. Probably the area between Tonopah and Goldfield in early 1900s. Collection of Ron Bonmarito.

A young boy operating a 'whim', or a rotary horse-powered well. In practice the whim acted like a winch to raise a bucket of water from the well. Location is believed to be in the Tonopah or Goldfield area. This is a rare photograph, as very few of these devices were used and even fewer photographed. Collection of Ron Bonmarito.

Where water was available, hydraulic mining was tried in Nevada. This is from Round Mountain, Nevada, about sixty miles north and east of Tonopah and situated at the foot of a ridge of mountains that obviously had enough water available to justify trying this type of mining. Dated sometime after 1907. Collection of Ron Bonmarito.

Tuscarora, Elko County, Nevada. The debris left after the hoist works exploded, probably a steam explosion. Collection of Ron Bonmarito.

An abandoned mine head and ore car, probably in the Goldfield area. Collection of Ron Bonmarito.

If you've ever wondered what happened to all those Model A Fords, this is one solution to the puzzle. This two-bit mine had an air compressor mounted on the back of a Model A and run off the drive shaft of the old car. Collection of Ron Bonmarito.

Not all mines were big operations. This is a small mine near the California-Nevada border. The nattily dressed man next to the lady is carry a small pistol on his right hip. Collection of Ron Bonmarito.

Arrowhead, Nevada was one of the many scams prepetrated on the public by unscrupulous land promoters in the early days. This may be the only known photograph of a town that never existed! Collection of Ron Bonmarito.

The town of Rhyolite, Nevada was established in 1905 and was gone by 1917. This photograph was taken in 1906 and shows four automobiles, all of them right-hand drive, it not yet being a law to have left-hand drive in the U.S. Collection of Ron Bonmarito.

The Carson River near Dayton, Nevada about 1890. These are the remains of the dredge that worked this river briefly in prior years. This dredge was not a continuous line bucket dredge but used a clamshell type of bucket to dig gravel from the bottom of the river and run it through a sluice on board. It was not succesful. Collection of Ron Bonmarito.

Rochester, Nevada was named by its founders for their home city of Rochester, New York in the late 1860s. It was not developed until about 1911, when promising silver ore was discovered there. It never amounted to much. Collection of Ron Bonmarito.

This is what may be the only known photograph of the town of Tobin, Nevada. Taken in 1908, it died almost as soon as it was founded. Collection of Ron Bonmarito.

Another classic boom town, Rosebud, Nevada in 1907. It lasted for a scant three years before folding. Collection of Ron Bonmarito.

Even Rosebud had a stock exchange. The unfinished building at left is the exchange. Temporary quarters are in front. Collection of Ron Bonmarito.

What might have been Jessop, Nevada lasted from 1908 until the end of 1909. Collection of Ron Bonmarito.

One of the pioneer miners in Nevada, George Lovelock. A studio portrait taken 5-11-89. Collection of Ron Bonmarito.

A later photograph taken by the Cat Tractor Co. in 1937 of a Tungsten Mine near Lovelock, Nevada. Collection of Ron Bonmarito.

A shift from the Ruby Hill Mine in Richmond, Nevada, dated 1881. The man seated at right front is named as a "Mr. Masterson." Many Wild West gunfighters put in time in Nevada mining towns. Collection of Ron Bonmarito.

The Weepah, Nevada business district in 1927. Available sources claim that Weepah had a brief boom in 1927 that lasted less than a year. The town post office was open for about 18 months. Collection of Ron Bonmarito.

Town site of Weepah with a couple of ladies in a new Chevrolet in 1927. Collection of Ron Bonmarito.

From the Kimberly mine in White Pine County, eastern Nevada. This was a high-grade copper mine. This photograph was taken sometime after 1905, when the mine was developed. All three men are using carbide miners' lamps. The size of the ore cart is much larger than those commonly used for gold or silver mining. Collection of Ron Bonmarito.

Kimberly Mine in White Pine County, Nevada. The man here is loading the cart, which is a side-dumping cart. Collection of Ron Bonmarito.

The town and mill in Tybo, Nevada in 1930, when it was a company town and the mines were operating rather marginally. The actual mines were up and down both sides of the canyon. None of these buildings exist today even though the ore bodies are not worked out. Collection of Alice Cleary.

The earlier mill in Tybo about the turn of the century, when mining was at a standstill. The mill is in bad repair and was eventually torn down. Collection of Alice Cleary.

The entrance to Tybo canyon. There is no hint of what exists beyond until you've entered the canyon, when the folds in the rock give a good indication of some very strange geology in the area. This road eventually leads to Belmont, Nevada, another ghost town. This photograph of a stage dates from the 1880s and is in the collection of the family which still owns property in the town. Collection of Alice Cleary.

CAFE

The town of Ruth, Nevada between 1915 and 1920. This is just the business district of the town, which was all company-owned—hence the absence of saloons. Collection of Ron Bonmarito.

Ruth & McGill

The last Nevada mining settlement that needs to be mentioned is the town of Ruth. Named for the daughter of a local mine owner, the town was established in 1903 to mine not gold or silver but copper. It was entirely a company town, established solely to house the miners. The ore body was determined to be very large and almost completely on the surface, thereby allowing the use of large earth-moving machinery, which by that time was being developed in earnest. Steam-powered shovels on rails were widely used, as were small steam trains designed to haul heavy loads up inclines to be processed in concentrators and then transferred to other railroads for transport to smelters.

The ore in Ruth was about 15% to 20% copper, with somewhat less than 1% gold and 1% silver. Rates of recovery were quite high, because effective new recovery processes were then coming into general use. The mines produced mountains of copper, with small amounts of silver and gold as by-products. All of it was refined at another town established some miles away, the mill town of McGill.

Since Ruth was a company-owned town, no bars were allowed and the town remained a quiet one for its entire life. All it did was produce lots of copper! At one time it had the world's largest open pit mine, which continued to grow even when it became necessary to move the town to get to the ore deposits underneath it. Since Ruth was owned wholly by the company, this was not a problem, and the original townsite of Ruth no longer exists. McGill, also a company-owned town, still exists as a slow spot in the road. However, the smelter is now gone, and with it the only real reason for the town's existence.

The company town of Ruth before it was moved to mine the copper underneath it. Collection of Ron Bonmarito.

One of the saddletank engines used in the copper pit at Ruth, Nevada. This is engine #338, an 0-6-0 oil burner. Collection of Ron Bonmarito.

The Ruth pit. Note the three men standing on the track on the lower left for an idea of the scale involved. Collection of Ron Bonmarito.

The Ruth pit. The loaded train of ore cars has three engines working to move it uphill. Collection of Ron Bonmarito.

Some of the drillers and powder men in the Ruth pit. Tons of explosive were used for each shot to break up the ore, which was then moved by large steam-powered shovels into the ore cars for movement to the smelter. Collection of Ron Bonmarito.

The copper smelter in McGill, Nevada, about forty miles away from the mines in Ruth. This photograph was taken after the operation was purchased by Kennecott. Collection of Ron Bonmarito.

The company town of McGill. These were sleeping shanties for the workers while the mill was under construction and the rest of the town was being built. About 1907. Collection of Ron Bonmarito.

Company town of McGill about 1912 after the more permanent structures were built. The Kennecott mill stood behind the photographer. Collection of Ron Bonmarito.

COLORADO

While the Comstock lode in Virginia City, Nevada was being developed, prospectors in other parts of the western mountains were hard at work digging also. Colorado was one of the earliest areas to be explored. People in the eastern population centers thought that Colorado was substantially easier to get to than California or Nevada, and in a few cases it was. Gold-seekers hoped to ride their wagons across the Great Plains from Missouri, and to turn either north or south when they hit the great ramparts of the Rockies. While it *was* easier to travel to Colorado than to other Gold Rush destinations, it wasn't as easy as easterners expected!

The earliest explorers were headed for the Pike's Peak region, because it was rumored that rich placer deposits of gold had been found there in 1857. Still, it was not until 1859 that serious deposits were found and the rush began in earnest. As an indicator of the level of activity for the year ending June 1st, 1861, the County Recorder of the Spanish Bar District noted 12,530 lode claims, 303 bar claims (placer claims on river sand bars), 25 gulch claims (placer claims in arroyos or dry watercourses), and 300 water mill sites.[10]

While there was certainly gold and silver here, it was a bit more difficult to reach than the easily worked placer deposits of the California gold fields. The early Colorado placer deposits were quickly worked out (depleted) and the lode claims were discovered. While the gold veins in California were quartz to a great depth, the Colorado veins were in the form of sulpherites that started just below the surface.[11] This meant that fundamentally different mining and milling practices had to be used, all of which required large amounts of capital for building plants capable of mining and treating the ore.

The problem was even more complex because this great, open area had no infrastructure to bring equipment to the mining areas. There was no railroad within a thousand miles, necessitating massive movements of men and materials by oxen trains over the plains.

Mules or burros hauling planks into the mines in Colorado. Because of the inaccesability of some of the mines in Colorado this was the only way to get wood, or anything else, into the mines. The caption reads "Some of Billy's Grass hoppers—Conrad Mines." Date unknown but probably 1870's. Collection of Ron Bonmarito.

A bird's eye view of the Elkton Mine in Cripple Creek, Colorado around the turn of the century. The blur at left is a moving train. Collection of Ron Bonmarito.

Blacksmiths at work swaging an axle or shaft using a swaging block and hammers. This method of hammering or compressing metal into a shape produced a very tough, uniform finish. From the Columbia Mine in Telluride, Colorado. Collection of Ron Bonmarito.

The richest of the ores came to be known as "tellurides" (hence the name of the town Telluride, Colorado). Tellurium, the element common to all tellurides, is the only known element which combines with gold in nature.[12] There are basically three different types of tellurides: sylvanite, which contains 30% gold, 10% silver, and 60% tellurium; calverite, which contains 39% gold, 3% silver, and 58% tellurium; and petzite, which contains 25% gold, 42% silver, and 33% tellurium. This does not compare well to most placer gold, which assays at about 90% gold and 10% silver.

Milling telluride ore presents unique difficulties. Though the problems were not unknown (they had been encountered in California), it was accepted that the very best of the stamp mills then in use could recover only about a quarter of the gold in the ore, and were incapable of saving the silver or the copper that occured in other types of ore. The silver and copper were found to be worth nearly as much as the gold when it finally became possible to recover these elements.

The boom towns in Colorado were every bit as wild and lawless as those in other areas, attracting the usual riffraff along with the serious miners looking for a better living. While Nevada was sufficiently remote to more or less preclude any effect from the Civil War, Colorado was a lot closer, and the various factions frequently fought for their chosen sides during the 1860s.

In the southern regions of the state, the San Juan mountains initially received little attention because of the hostile Ute Indians who claimed the area as theirs. By 1873 the Utes had signed a treaty (for "cash consideration") and the rush was on. Some famous names come out of this area: Telluride, Ironton, Red Mountain, Ouray, and Silverton, to name a few. The ores were largely silver in content but enough gold was found in places like the Camp Bird Mine to extend the boom into the twentieth century. Because most of the mines were located in very high mountains (some as high as fifteen thousand feet), supplies and men were generally brought in on pack trains of burros.

An early photograph of the town of Telluride, Colorado. The mountains in the background give some indication of the problems associated with mining in this area. Collection of Ron Bonmarito.

A very early photograph of a Telluride, Colorado mine. Notice the total lack of vegetation caused by the need for wood for the mine and firewood for everybody. Collection of Ron Bonmarito.

SOUTH DAKOTA: THE HOMESTAKE MINE

The history of the gold strike in the Black Hills of South Dakota is well known. For some time prospectors had known that gold was to be found there, perhaps from the Indians who considered the hills sacred or from early white traders who may have found gold there. In any event, the great Homestake strike and rush started during the 1870s, and the lode has been continuously mined ever since. While not strictly speaking part of the West covered by this book, the Homestake Mine in South Dakota should be discussed because it is so significant in size and longevity.

The Homestake Mine of today is a large conglomerate which was originally owned and controlled by the Hearsts of Hearst Newspaper fame. Hearst was also an early player in the Ophir Mine of Virginia City, Nevada. The Homestake started out as a series of smaller mines which were gradually bought up to bring all the various ore bodies under a single ownership.

The ores of the Homestake were not the rich quartz veins of California or the tellurides of Colorado. Rather, they have been described as consolidated gravels which over the eons became conglomerate rocks, containing particles of gold usually too small for the unaided human eye to find. These ores needed to be moved in quantity, as they yielded only $3 to $6 per ton. Compare this to the $3800 per ton of the Comstock Mines, and you can quickly see the massive job involved in making the Homestake mine pay. But pay it has, because of the utter massiveness of the ore body underlying the Black Hills.

The Homestead strike was an archtypical boom. Consider the town of Custer in 1876; when news arrived of the strikes at Deadwood, South Dakota, Custer was estimated to have about ten thousand people in it. In less than a week Custer had been almost completely depopulated; in fact, one authority stated that only fourteen people remained after the rush started.[13] The town of Deadwood was laid out on April 26, 1876, and by the end of summer it was occupied by an estimated 25,000 people. It became one of the wildest places on earth, the town where Wild Bill Hickok died in one of many saloons.

Two brothers, Fred and Moses Manuel, are generally credited with the initial discovery of the Homestake Lode in the spring of 1876. It is interesting to note that the first type of rock crusher to be used there was the old reliable arrastra. It was not long befire a syndicate led by Hearst stepped up to purchase controlling interest in the mine, providing the capital needed to develop the ore body. Their confidence was well placed, as the Homestake became a dividend-producer almost from the start. By 1880 some $450,000 in dividends had been paid to stockholders. While this number varied over the subsequent years, it never failed to produce bullion and dividends.[14] By 1900, dividends had reached $1,175,000 per year.

This is truly amazing considering the amount of rock that had to be moved and treated in a very deep mine. In 1902-1903 the ore averaged $3.539 per ton.[15] To process all this rock Homestake had a thousand stamps working in its various mills and two of the largest cyanide treatment plants in the world.[16] The miners and management very quickly adopted the square-set timbering system which Deidesheimer had developed for Hearst's mine in the Comstock Lode, the Ophir. This system was adapted to the ore body in the Homestake, and some innovative filling and shoring techniques allowed the company to extract great sections of rock to mine the low-grade ore.

At first, all this moving of rock was done entirely by hand, with real divisions of authority and work established by the company. Muckers, the lowest of the low, were required to load sixteen tons of rock during a shift. If a man proved capable of doing this regularly he would be elevated to the status of miner. Miners used hand tools to drill the rock so the powder men could come in during the shift changes and blast. Eventually (in 1894) this system was augmented by compressed air drills, which greatly speeded up the processes of drilling, blasting, and ore production.

The Homestake was (and still is) a well-run operation, and is the first to use new techniques and equipment. Because of this approach and because of a very good working relationship between management and labor, the mine continues to be a viable operation.

MONTANA

These two intrepid prospectors are in a town in Montana heading out for the gold fields. One of them has his bedroll over his shoulder while the burro has everything else. These do not look like experienced miners but they are certainly typical of many men who traveled to the west to seek their fortune. Collection of Ron Bonmarito.

The search for gold in the west continued into the northern reaches of the country. Montana's first recorded gold find was in 1852 in an area west of what is now Butte, Montana. Since the area was very remote and peopled primarily with hostile Indians, little was done until late 1862 or 1863. The documentation of exactly who found what and when is somewhat garbled, but it is generally agreed that a man by the name of John White and his party found a good placer claim on July 28, 1862 at a place they called Grasshopper Creek.[17] The camp here eventually became known as the town of Bannack, and commercial gold mining began in earnest. While mining here was in relatively good ground (productive and easily worked), this richness apparently did not extend into other areas to the same degree.

This area is probably as well known for the lawlessness of its communities as it is for the gold taken out. Among the lawless of the region was the infamous Henry Plummer. Plummer, known as "Soapy Smith" of Montana, was the head of a very organized gang of notorious thieves and robbers who ran all sorts of illegal operations. The members of his gang were known to be very dangerous, killing without mercy. Henry Plummer was hanged by vigilantees in 1864.

A good indicator of the rough-hewn nature of the camps is a letter dated April 30, 1863, from a lady named Mrs Emily R. Meredith. In it she wrote about the price of flour and the good pay from the claims. She also discussed the hellish nature of the community; she wasn't sure how many men were killed the previous winter, but wrote "[the fact] that there have not been twice as many is entirely owing to the fact that drunken men do not shoot well."[18]

At the same time, in what came to be called Virginia City, Montana (not to be confused with the city by the same name in Nevada), a rich placer strike was founded in a spot known as Alder Gulch. Some of the claims here were very rich, and some serious lode claims were later filed. In the space of a few years some $30 million was taken from this area, and it eventually yielded about $85,000,000.

This was tough country and hangings were common even after the Plummer Gang was killed or run off. Just getting to Alder Gulch was one of the more difficult journeys for people coming from the East.

To get to Alder Gulch, prospectors had to travel a route that came to be known as the "Bloody Bozeman Trail." The Bloody Bozeman Trail cut north and west from the Emigrant Trail, which continued on through Wyoming on its way further west. If a gold seeker kept to the Emigrant Trail he could get to the gold fields in Montana eventually, but it was a long, difficult road that crossed the Continental Divide twice. By taking the Bozeman Trail travelers avoided this arduous trek; in the process, however, they exposed themselves to attacks from the Crow and Sioux Indians, both ferocious warrior nations who did not like white men. Add to this the "Road Agents"—thugs who preyed on people returning from the gold fields—and you had the makin's for some real problems!

Although stories told in print and on the silver screen portray the trip west as a hair-raising adventure for every wagon train, there seem to have been plenty of people who made it through to Bannack and Virginia City without so much as a shot being fired. Even the appointed governor of the territory of Montana made it there without a scratch.

As in other mining camps before the laws of 1872, what constituted a "claim" was determined by common consent in each camp, or by a majority rule. At Alder Gulch, a claim was initially defined as a hundred feet of the creek, running from rimrock (the top of the bank) to the opposide rimrock. When latecomers objected that it was done differently in California, the rules were changed to measure the claims from the center line of the creek to the rimrock on one side. Since it was often ambiguous where the center line of the creek was, discord was sure to follow. There was reason to be greedy, as the gold here was plentiful and very pure, though sometimes very deep below the surface. One assessment valued the ore at $18.50 per troy ounce; compare this to the $16 assessment from Idaho and $13.50 from Butte.[19] The difference in value comes from the amount of silver, copper, and other impurities in the placer gold.

As in any mining boom area, prospectors reasoned that there should be more gold in the outlying areas. True to form, they did manage to find other pay streaks outside of Alder Gulch that yielded silver and gold ore, which was sometimes difficult to extract.

Butte, Montana

The town of Butte is another example of a boomtown, mining not gold or silver this time but *copper.* The town's economy had been very unstable for years, even before the silver panic of 1893 almost did it in for good. While apparently it had always been known that copper existed in Butte, that source of profit had been overlooked because of the difficulty of smelting the ore; smelting copper gives off highly toxic fumes that in one incident actually killed fifteen people. In 1884 a smelter was located away from the town by a group of partners who were incorporated into the Anaconda Mining Company in 1894.

The story of Butte, Montana is well worth telling, a rich collage of the usual comings and goings of people in mining camps. The city of Butte was populated primarily with Irishmen, many of whom had worked in the original Comstock mines and had learned their trade there. Another substantial number of Irish miners came directly to Butte for the work and for the community of their countrymen. Butte was unique, because of all the towns in the western United States, it became most like European factory cities in this time. Butte was genuinely ugly, largely as a result of the copper smelters belching forth huge quantities of acrid smoke and dust.

This was a city of men; only the Irish men made the journey out west. Wives, if they had them (and more likely they did not), were still back on the old sod. These men needed release from the mental strain of working underground seven days a week for months on end, and as a result, a tremendous red light district developed. It was reported that Butte's district was surpassed in size only by the one in New Orleans.

The Union Pacific Rail Road extended to Butte in 1881, making transportation in and out much easier. Butte's miners were well paid for their time, at about $3.50 per day—a little less than the Comstock miners but still above average. The city continued to prosper, producing copper by the ton. As in most mines of this type, by-products of the copper mining were silver and gold, and in Butte the silver provided pure profit for the owners. Today Butte is a modern city, still nurtured by the copper ores of the Anaconda and other company mines.

Montana is a large state, and it took some time to find other gold-bearing claims. Places named Diamond City, Confederate Gulch, Montana Bar, and Helena all produced gold in commercial quantities. Montana Bar was called "the richest half-acre of ground in the world" when four miners took out some three and a half tons(!) of placer gold during one season.[20] Despite this promising beginning, however, gold mining came to end there when the placer gold was gone; there were no large underground lodes to take over.

Butte, Montana

A good illustrative photograph of Butte, Montana around the turn of the century. The photograph is not washed out; what you are seeing is the incredible air pollution that existed there from the copper smelters. Note the large piles of stacked firewood alongside what appears to be a train loaded with more wood. Collection of Ron Bonmarito.

OREGON

Geologically young and active Oregon was one of the earliest areas to be settled by westward-traveling Americans. Settlers were moving to the state before the time of the Sutter's Mill strike in California, and they were planning to develop agriculture, not gold mining. There are accounts of gold discoveries in various places during the earliest days of Oregon's settling—even earlier than the discoveries in California—but no really big search for gold started until the strike to the south took place in California. The lodes found in Oregon were originally thought to be extensions of those found in the Sierra Nevada range in neighboring California. Eventually some major mining took place in the state.

There is some dispute as to whether some of the mining took place in Oregon or in California, since the borders of the two states were not well-definined in those early years. Some modern writers have taken the view that a significant amount of the gold reported as having come to the San Francisco Mint from northern California actually came from mines in Oregon. Documentation to support both sides is available, and the government records of the day are somewhat sketchy.

There is also the distinct possibility that large amounts of gold were taken out of the area by Chinese laborers who mined tailings and low-grade deposits in California, Oregon, and later Nevada, taking the unrefined gold directly back to China with them. The Chinese were not allowed to own a mining claim until white people had abandoned them, but even so there were large numbers of Chinese in the diggings from the earliest days. There is an almost total lack of written information about their successes in the recovery of gold. This subject has the aura of a mystery about it.

In Oregon, no one well-defined ore body could be identified, even though there were some individually rich placer and lode claims. Mining in one form or other started as early as the 1840s, continuing spasmodically for the next several decades.

The real 'gold strike' for Oregon residents was agricultural, as produce from Oregon's agricultural fields was transported by steamer to be sold in San Francisco, and thence into the California gold fields to feed the miners. The Oregon 'gold rush' to sell foodstuffs and lumber in San Francisco began in earnest after the first Oregonians came back from California with their pockets full of freshly-dug gold from the Mother Lode's riches.

Although gold had been found (and lost) earlier in Oregon, the real mining rush there began sometime during late 1851, when prospectors found rich placer deposits in the south of the state, near what would become the town of Jacksonville. This is in an area close to the present-day boundary of California and Oregon. The miners were not sure which state they were in—so when the tax collectors came around they claimed to be in neither, and paid taxes to nobody.

Oregon towns with names like Sumpter, Galena, Susanville, Placer, Greenback, and Golden all produced gold and silver. They also produced cinnabar, which was used for the production of mercury, which in turn was used extensively in the recovery of gold. The stories surrounding these and many other towns are similar to those from California and other boomtown areas. Gold was found in rough deposits that soon played out, but not before some impressive numbers were accumulated. Surprisingly, in this part of Oregon (southwestern, but inland from the coastal mountains), water was a problem for the placer miners. In some cases good quantities of water did not become available until the late 1870s, when ditches were dug from the larger rivers to supply the mining sites.

In a few places gold was found on the ocean beaches, where it had been washed by rivers in times past. While these beaches were prospected up and down the coast, they were not large finds.

Other towns like Grants Pass, Prineville, and Bend survive today because they are transportation points, agribusiness centers, or recreation areas beckoning city dwellers to spend 'paper gold' during their preferred seasons.

Topographically most of the western part of Oregon is much like the coastal region of California, with major mountain ranges rising steeply out of the Pacific. Beyond these ramparts, the interior and western areas of the state are semi-arid basin and range sections much like those of central and eastern Nevada.

The richest placer claim to be discovered in Oregon came from this rugged, dry country. In 1862, at a place that came to be known as Canyon City, a placer claim was filed that by 1901 had produced about $26,000,000 in gold.

There were both placer and lode mines in this area. Eventually dredges came into use, and continued to be used until the early years of this century. These extracted good amounts of placer gold from the sands of the John Day River bed.

Near the present day town of Baker, Oregon is the Virtue Mining District, so-called because most of the gold taken came from one mine, the Virtue Mine (named after a man with the last name of Virtue, by the way). It operated from 1862 to sometime in 1924 and had some very rich ground. The Virtue was the major producer of the district and is credited with over $2,000,000 in production during its lifespan. Being a lode claim it had chemically-free-milling white quartz ore. With this type of ore, the gold could be easily separated from the quartz matrix by mechanically crushing it, and then running it through a sluice box system. The gold needed no further reduction by any other method before being smelted into bars.

Some of the other mines in the area were placer mines, which worked the ground outside of the boundries of the Virtue.

West of Baker was another area where some mining took place. Like other mines, it was first discovered by placer miners during the 1860s. Sumpter was a sort of hub for the area and eventually became the home of a large stamp mill to process ore taken from the hard-rock mines then being developed. Though the town had a post office by 1874 (a sure sign that some people intended to stay there for a while), the major growth did not take place until the late 1890s.

Much of this development became possible largely because of the advent of the railroads, which made it possible to bring in supplies (and people!) and take out ore for smelting in already existing mills in other areas. The Oregon mines, while rich, did not approach the production of the California or Nevada mines and many were not well developed until later times.

The reputation of the Oregon mines was tarnished by a couple of major fraud cases during the decades immediately following the turn of the century. This, coupled with the Great War, ultimately spelled doom to the mines; investors were unwilling to speculate their fortunes on them, and did not consider them viable businesses.

At the same time, directly to the south, Nevada was still developing new fields for silver and gold production, as were other states in the West.

The granite Hill Mine near Grants Pass, Oregon was one of the few gold and silver mines in this area. The large angular building in the center is the mill. Collection of Ron Bonmarito.

WASHINGTON

Gold mining in Washington came about as a result of the general rush west to the gold fields of California in the 1850s. In general, Washington played only a small part in the production of gold and silver until the rushes to the Yukon in 1898 and to Alaska and Canada later. Then, the port of Seattle became the jumping-off point for all north-bound gold seekers.

Gold was discovered in Washington itself for the first time in 1855, near Fort Colville, an old fur-trading post near the Canadian border in the northeastern corner of the state. Though the mining here was relatively short-lived, the area was suitable for agriculture so most of the prospectors simply stayed on to farm the land.

In 1886, along a stream called Salmon Creek (every northwest state and county seems to have a Salmon Creek!), a good placer find was recorded and Salmon City was founded. Since the area had an abundant water supply, the placer deposits could be worked easily. Soon the town of Ruby became the center of mining in the area. Some good-sized silver claims were found and worked, but the depression of 1893 (a result of the silver-market crash of the same year) effectively put an end to the Ruby Mining District.

In another area of the state, the Eureka District, the Republic Mine set off a major rush which ended about as soon as it started. Even though many claims were filed and a few mines were actually operated, the ore was not rich enough to sustain any major long-term mining. Though initially the Eureka District had some good prospects, the quantities of raw ore were just not there.

In central Washington several locations showed promise with placer claims, but the big lode strike never did occur. The town of Liberty was one such location, producing pockets of placer gold and nuggets for miners to drool over. With no lode mines to sustain long-term mining, however, the area was bypassed quickly. In the twentieth century a dredge was brought in, and it successfully produced gold for a few years. Down the road, the town of Blewett was established, circa 1861. While there is some debate about when things happened in Blewett, a stamp mill was built around 1879, and it did process gold ore. Beyond that, little is definitely known, since a disastrous fire destroyed most of the town and it was not rebuilt. Now a highway runs through—or rather over—what had once been Blewett.

The real 'gold mines' for Washington towns were agricultural, supplying goods to the more productive mining areas to the south and east. Another 'gold mine' was the abundant timber.

The first outcroppings and workings of the Republic Mine in Republic, Washington. The timbers were as much to keep the rock walls from collapsing as they were to help climb down into the ore face. Collection of Ron Bonmarito.

RALSTON & HAMMOND'S

MAP

OF

Republic Camp,

EUREKA MINING DISTRICT.

COMPLIMENTS OF

Graham & Garrett,

AND

J. W. Heisner & Co.

Miners and Brokers.

Mines and Stocks for sale in British Columbia, Idaho, Okanogan and Republic Camps, Wash.

Confidential Reports on Mines anywhere in the West.

Spokane, Wash., 211-212 Peyton Bldg. Republic, Wash., 305 Clark Avenue.

Address all Correspondence to 211-212 Peyton Block, Spokane, Wash.

Bank References: Exchange National Bank. Republic Bank: Fidelity National Bank.

Left:
These fold-out maps were given away by promoters of land and stocks. They were virtually useless for anything else. Shown about actual size. Collection of Ron Bonmarito.

375325

MAP

OF

Republic Camp

EUREKA MINING DISTRICT.

COMPLIMENTS OF

C. O'Brien Reddin & Co.

Miners and Brokers.

MINES AND STOCKS FOR SALE

In British Columbia and Republic Camp, Washington.

Confidential Reports on Mines Anywhere in the West.

SPOKANE, WASH. *514-515 Peyton Blk. 'Phone Main 25.* ROSSLAND, B. C. *13 Columbia Avenue, 'Phone 68.*

Right:
Another map of the Republic Camp mining area from another promoter. Collection of Ron Bonmarito.

WYOMING

The South Pass area of Wyoming is a high plains pass, which most people went through while traveling overland to Oregon or to the gold fields of California. While it was the easiest path over the Continental Divide of the Rocky Mountains, it was also a spot that a few people chose to search for gold once it became known that gold was out there for the taking. Unfortunately, it is fairly high in elevation. As some people came to know, that made it dry in the summer and brutally cold in the winter.

Such is the lure of gold that men will endure almost anything to pull a living out of the ground. The towns that sprouted there are almost all gone now, with remnants of mines slowly disappearing into the hills. Places such as Atlantic City, South Pass City, Miners Delight, and Pacific Springs all had their days of glory. Typically, they started with the discovery of placer deposits that proved difficult to mine because of the lack of good water. Soon, the ledges in the area were prospected. They were found to contain free-milling quartz/gold ore, much to the delight of the miners (hence the name of the one town).

By 1867 lode claims were being filed, and by 1869 the population of Atlantic City had risen to over two thousand people, including women. They conducted all shaft and drift mining to follow the vein know as the Atlantic Ledge. The location of the area was about one hundred miles north of the brand new Union Pacific Railroad, allowing stamp mills to be brought into the area at the earliest possible time.

Unfortunately, like many other gold fields, the Atlantic City ledge was pretty well mined out by 1875, although a dredge operated in Rock Creek Canyon in the early 1900s. The towns slowly died and today are of little note except for South Pass City.

South Pass City is a real exception to the tradition of mining boom towns. Its story is worth telling briefly. Founded in 1867, it shortly achieved a population of over two thousand hardy souls. By all accounts, it was a family community complete with wives and children.

The fact that the area was subject to occasional hostile Indian raids in those days probably had a lot to do with the cohesiveness of the community. The women were well-organized and succeeded in introducing and passing a bill giving the vote to women in Wyoming. A certain lady by the name of Mrs. Esther Hobart Morris was shortly afterward appointed the local Justice of the Peace and served for many years in that capacity. The town had the first bank in the area, as well as the usual number of saloons, a school system, a newspaper, and regular stage service to other towns in the area.

The Carissa Mine just outside of town was the biggest producer in the area. When its vein pinched out so did the life of the town. By the early 1880s South Pass was nearly a ghost town.

Eventually Wyoming's real contribution was providing transportation stopover points and fertile cattle ranches. Much later, oil was discovered. Coal is still being mined in abundance.

UTAH

Officially, the earliest discoveries of precious ore in Utah were in 1863. It is known that some silver mining was being practiced in the state much earlier by both Navajo and Spanish-speaking peoples. The early history of the state is a troubled one because of the antagonism that existed between the Mormon Church (and therefore the Mormon-operated Utah Territorial government) and the United States government. U.S. Army troops were stationed in the Salt Lake area as early as 1857 to "keep the peace."

The commanding officer, General Patrick E. Connor, encouraged his troops to prospect in the mountains. They uncovered silver ore in Bingham Canyon in September of 1863. Claims were quickly filed in the name of the Jordon Silver Mining Company, which had twenty-five stockholders. Good-sized ore bodies were discovered, mainly of Galena-type ore comprised of lead, silver, and zinc. This can be difficult ore to mill and smelt, so one of the first tasks was to build mills and smelters.

During the same year claims were filed in the Stockton region, and near Salt Lake the Little Cottonwood District produced over 3500 claims in a short time. These mines were mostly lead, silver, and zinc mines, with virtually no placer mines producing gold. This district's good years were quite short-lived, even though the mines were reworked in later years.

One of the exceptions to this trend was in the Tintic district, which, like others in the state, was in mountainous terrain that was difficult to access with the transportation of the day. The Tintic district was discovered in late 1869, and was organized in the spring of 1870. The first claim filed was the Sunbeam on December 13, 1869, quickly followed by the Black Dragon, the Mammoth, and the Eureka Hill claims. The Mammoth Hill was the biggest producer of this group and accounts today for the existence of Mammoth, Utah.

Initial development was slowed because of the poor facilities for bringing in men and materials and for transporting ore out for smelting. The earliest ores produced were mined off the surface and were high-grade enough to warrant shipping them out by pack train and freight wagons. Some were shipped as far as Swansea, Wales for smelting, while others went to San Francisco, Reno, and Baltimore.

It was obvious that local mills and smelters needed to be built. The first mill was built in Homansville in May of 1871, with a second to follow in September of the same year. Others followed in 1873, but the smelters generally were not successful because the ore was difficult to separate. These were complex ores that had many minerals in them other than lead, silver, and zinc. Some of them contained small quantities of gold as well as bismuth and sulpher.

Leaching the ore was tried as a method of reducing it but this was discarded very quickly. The Old Millers Mill in 1879 was opened and closed almost in the same year. One of the problems associated with milling ore in this area was the scarcity of water for use in amalgamation pans. The problem was not solved until some years later when a reservoir and pipe system was put into use. Until that time much of the production of the mines was shipped to mills in the Salt Lake area.

The Tintic district was a good producer of silver, lead, copper, and some gold, but figures for production up until 1880 are not available. It is estimated that production during this time period (about ten years) was less than $2 million total. From 1880 to 1898 gold production in ounces is given as 277,142 while silver production is much greater at 34,575,199.[21] Lead and later copper production was just as high.

Geographically, the Tintic district is located about dead center in the state, with arid mountains predominating. Production was greatly enhanced when two railroads were pushed through, one from the west and one from the east. The first to reach the district was the Oregon Short Line in 1883. The Rio Grande Western came in from the east in 1891.

Some famous Utah places were originally mining centers, like Alta and Park City, now both major ski resorts. Alta was the location of one of the early discoveries of high-grade silver ore. Unfortunately the initial mines were pinched out by faulting in the mountain slopes; that is, the ore veins came to abrupt stops where fault lines had shifted the rock and broken the veins' continuity. It was sometimes possible to expand a tunnel and find the continuation of the vein elsewhere, but usually the effort was fruitless. While some serious production did take place in Alta, by the 1890s the town was almost dead.

Interestingly, while the Mormons decried the presence of "ladies of the evening" and the prodigious consumption of intoxicating liquors, they didn't run these practitioners off. The actual Mormons among the miners were apparently few in number because Brigham Young urged his flocks to raise corn and vegetables in abundance. After all, somebody had to feed all those miners!

One last item of note is the Town of Corinne, Utah. Not a mining town, it is worth mentioning because it was the location of several smelters to process the gold and silver ores from the mines to the south. The streets of Corinne were originally paved with the slag from the first smelters. Some intelligent soul reasoned that since not all of the gold and silver was recovered by the smelters, the streets should be tested for metal content. Sure enough, they contained enough metal to make it worth while digging them up! They did, and ran the whole town back through the smelters to recover the gold from the town with streets *literally* paved with gold.

IDAHO

One contemporary account describes the sands of Idaho as having gold in the gravel wherever it was found, even underneath lava flows. That's probably an exaggeration, but in the decades following the California and Comstock strikes, northern Idaho became a transit point for travelers going to Washington and Oregon from Montana and back. The first documentable prospecting took place around 1878 in the Coeur d'Alenes—surprising, since the area had been surveyed as early as 1860 for railroad rights of way.

This is exceptionally rugged country, and the placer gold found there initially was rough and high grade. While early claims yielded enough gold to start the rush in 1878 and 1879, the biggest claims (in silver) were not discovered until between 1882 and 1884. The Northern Pacific Railroad had been working on its roadways during this time and had finally achieved a connection on September 8, 1883. True to form, the railroad promptly put out fantastic claims of gold strikes which generated hoards of gold seekers—all riding the railroads at good fares, of course!

As usual with boom areas, prospectors went in all directions. Soon the North Fork of the Coeur d'Alene, the site of the early strikes, was outstripped by the really big claims of the South Fork. These claims included the famous Bunker Hill claim and the Sullivan claim. While the North Fork claims were placer and gold lode claims, those of the South Fork were mostly in galena ore, which is a heavy lead and silver ore. At first the ores were not processed there, though transportation of the ores from the mines was difficult at best. In some cases this involved shipping ore first by wagon over incredibly bad roads to Coeur d'Alene Lake, then by steamer to Coeur d'Alene City, where it was put on the railroad for treatment in Wickes, Montana.

The problem became one of economics; how could the mines continue to operate with no mills close by and no efficient way to move the ores to existing mills? The answer was to build more railroads throughout the area and ship lower grade ores in bulk. Of course, the railroads were only too happy to do this for paying customers. The Northern Pacific, the Union Pacific, and their smaller connecting lines became wealthy from the outflow of the prolific mines of the areas.

Eventually even this high quantity shipping became too expensive, as the ore encountered became a less-profitable lower grade. To avoid transportation costs, the mines built their own mills and concentrators in close proximity. This was a gradual process; only a few years before the 1920s was it finally accomplished.

In Gem, Idaho this is the Frisco Mill on July 11, 1892, the day after the mill was blown up in a labor dispute. Idaho mines had some real union battles during this time. Collection of Ron Bonmarito.

Idaho had some decent-sized mining operations. This one was the Morning Mine and Mill in Mullan Idaho in 1893. Collection of Ron Bonmarito.

ARIZONA

Mining in Arizona for gold and silver is as much a story of phantoms as it is of reality. Tales abound of gold and silver mines lost and never found. The fables start from the earliest times in the mid-1500s, and continue on to the present. But for most part these are nothing but tall tales.

Gold placers did appear in a few spots in Arizona, and eventually some very high grade silver deposits were found. This occured primarily after the Spaniards had left, when the 'Californios' made their way into Arizona after their own California gold fields had played out.

One of the obstacles common in gold placer mining was the need for water, which is often completely absent in gold deposits located in dry streambeds. This was a severe handicap, since the most common method for mining placer deposits was using sluice boxes and gold pans, both requiring water. In the absence of water, a method of dry-washing was used that involved tossing gold-bearing dust into the wind and letting the lighter sand blow away, much like the winnowing of grain and chaff. This was not a very good method to recover a lot of gold, but it did work to a degree. The desert miners used it in Gila City, where one of the earliest placer gold discoveries was made in 1858.

When gold was found on the banks of the Colorado in 1862, the stampede was on in earnest. The town of Wickenburg was founded because of the location of a high-grade ledge of gold-bearing sand that came to be known as the Vulture Mine. Prospectors quickly came to establish claims for silver and copper, almost as an afterthought, when it was found that there were no extensive gold deposits.

Silver and copper, often found together, were found in abundance in several places in Arizona, despite attacks by indigenous Indian tribes who greatly resented the intrusion of the white prospectors on their land. By the middle and late 1870s several large silver discoveries had been made, and mills were developed to process the ore. In 1877 the town of Tombstone was the location of a major silver discovery, which eventually produced

The Little Coronado Mine in Metcalf, Arizona about 1900. Note the ruggedness of the terrain. This kind of country delayed development of mining in this area. Collection of Ron Bonmarito.

some $50 million in silver in the short span of five or six years. That the town did not die is something of a mystery. Today Tombstone lives on as the "town too tough to die."

Throughout the state of Arizona there were perhaps a dozen or so areas that produced gold and some silver. Shortly after the active demise of the gold placers, however, miners began trying to develop the copper deposits that existed in abundance. There were about twenty areas with copper ore of sufficient quality to justify raising money to commercially produce the metal. With the wider use of electricity and the need for more and more copper wire, copper became king in Arizona.

By the late 1880s, annual gold and silver production had dropped to about $4 million in Arizona, while copper production exceeded that value. In later years copper production yielded tremendous income to the mines of Arizona. By the turn of the century copper had completely surpassed gold and silver production. Today there are virtually no gold or silver mines in the state.

Mining copper is purely a question of capital. It takes a lot of money to produce copper because the ore must be moved out of open pit mines in huge quantities, then processed in mills requiring large buildings. Then the finished products must be transported by railroad. This is not a business for the 'little guy'. The copper industry is dominated by a few very large companies.

Interestingly, more gold and silver were mined (as copper by-products) during the period of heavy copper-mining than had been mined during the early years, when miners had ignored copper ores. Copper ore consists generally of 17% copper, 0.5% gold, and 2 or 3% silver. As copper mining expanded beyond all expectations, the incidental production of gold and silver became significant. One ton of copper ore might yield 340 lbs. of copper (worth $200), 20 lbs. of silver (worth $400), and 2 lbs. of gold (worth $3200). The numbers added up!

The actual entrance to the Little Coronado Mine in Metcalf, Arizona. The man standing in the open entrance gives some idea of the scale involved. Collection of Ron Bonmarito.

Little Coronado Mine opening. The men are sacking high grade ore for hauling out on the backs of burros. Collection of Ron Bonmarito.

The burro train hauling lumber into the Little Coronado mine in Metcalf, Arizona. Everything coming into or out of the mine traveled this way. Collection of Ron Bonmarito.

The cook shack at the Little Coronado Mine in Metcalf, Arizona. Notice the Chinese cook.
Collection of Ron Bonmarito.

MINING EQUIPMENT & MEMORABILIA

Like many other industries during the nineteenth century, mining began with the simplest of hand tools and chemicals, progressing to more and better tools, and eventually to whole new ideas about how to get gold out of the ground. Still, it is the simple tools that generation after generation of miners continued to use. The common pick and shovel were standard tools for all prospectors. They varied little in design but were all intended to move rock, sand, or gravel into a sluice box—or at its most fundamental, into a gold pan.

Gold pans as we know of them today are rather larger than those the early '49s used, since these men primarily used small sheet-metal plates with a slightly raised edge. With this one simple tool it was possible to wash gold out of a creek all day, cook your ration of beans for the evening meal, and then (simply washing the pan out in the creek or sand bar) get back to work. Sheet metal skillets were easy to use, but in a pinch cast iron cooking utensils could be used to pan for gold as well.

Shovels could be made out of a single piece or several pieces of sheet iron or steel, and many were simply forged out of the basic materials by local blacksmiths. These generally proved unsatisfactory because of the shape and quality of materials. Miners bought factory-made pieces as soon as they became available to them. Shovels of that time period tended to be rather flatter in shape than those currently made, and had distinctivly shaped shanks for the fitting of handles, which were very straight. It was also common for early shovels to be made in two or more pieces, with the blade rivetted to the shank. This is readily apparent when you see such a shovel for the first time. It is also the reason these shovels wore out faster and were discarded in favor of single piece stamped steel shovels when they became available. Finding a genuinely old shovel in any kind of usuable shape today is a rarity, since most of them have long since been eaten up by rust and simple usage.

It didn't take long for hardware manufacturers to come up with a better-shaped gold pan, and when they arrived at the optimum design they kept it. The gold pan has changed little in design since the earliest days and only the practiced eye is now able to tell the difference from true antiques and pans of current manufacture. Gold pans came in a variety of shapes and sizes and most will be found rusted out with age and use.

Picks and hammers (sledges) were used by everyone and are quite common to find. They can be distinguished most of the time by the shape and wear patterns, and less often by any casting marks that can be seen. The collector of mining hardware should beware of the 'Abe Lincoln syndrome': "That axe was Abe's, but it has had four new heads and six new handles since then."

About 1890 and the candlestick in the rock is showing off the ore above the man at left. This is a good illustration of how the candlestick was used. Collection of Ron Bonmarito.

These three miners are actually from Eureka, Nevada having their picture taken either in San Francisco or by a wandering photographer in Eureka. The chief items of interest here are the oil lamps on the hats of the two miners. These are very rare today. Collection of Ron Bonmarito.

When today's buyers of mining hardware look for bigger pieces, they can find a bewildering variety of larger objects. Items such as ore carts and dredge buckets are good examples, and can bring some rather princely prices in today's market. Ore carts, for instance, are available more in the western United States than elsewhere because they are rather heavy; they do not lend themselves to being mailed home from a Nevada vacation! Many of these carts were manufactured at foundries in the west and will be found most often near the areas where they were used. There is some variety in them because of how they were used but the overall design is of a very heavy undercarriage topped with a sheet metal box which was usually rivetted together by handset rivets. The early boxes were never welded with electric or gas welding, since those systems had simply not been invented yet. Wheels were castings that could be finished by filing or machining and were generally very finely made. Occasionally wheels were built with a center of wood bound with an iron strap or tire. These are rare to find today since they usually wore out and were discarded or rebuilt. Most of the carts were designed to be hooked together in a train which could be drawn by man or beast. It is not uncommon to see carts of different sizes and manufacture in use at they same mine, because they were bought and sold, moved around the country as needed, and generally traded like we trade in cars today. When the tops wore out they were rebuilt. They were designed to have this done on a regular basis.

Dredge buckets are another area of collecting interest but they, like ore cars, have practical limitations caused by their size and weight. These buckets come in different sizes depending on the size of the dredge and its capacity to handle pay dirt. The buckets are sized according to capacity: how many yards of material they could carry. They are all very heavy cast iron with replaceable edges, which are riveted in place using handset rivets. In most cases, moving one of these buckets is equal to lifting the motor out of your favorite car. They're heavy!

The ultimate collectible would be your own dredge, but we're aware of only one of those in the world, located outside of Fairbanks, Alaska. By the way—it isn't for sale.

An electric dynamite cap charger, this device produced enough electric current to set off the explosive caps which set off the main charge. Collection of Ron Bonmarito.

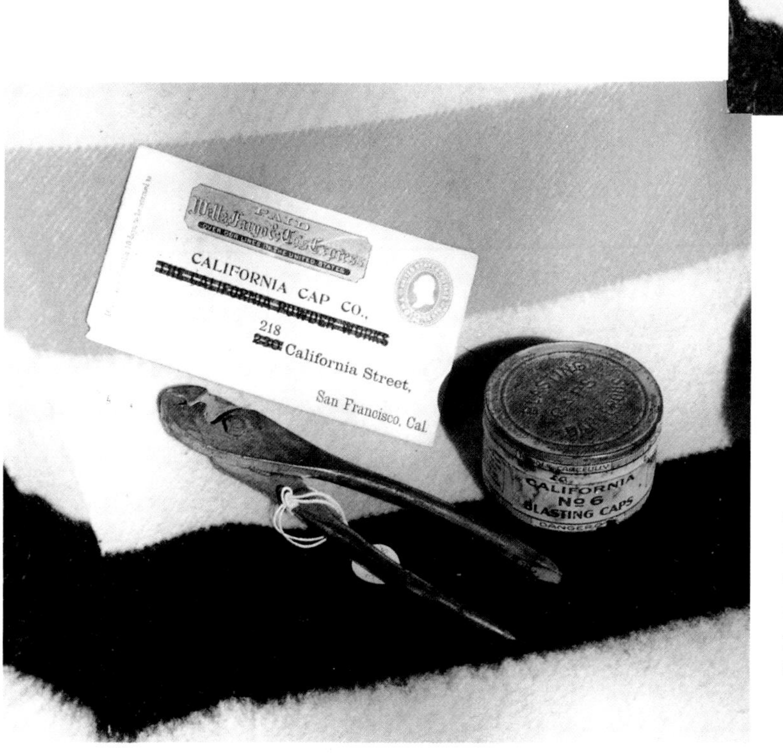

Tools of the powder man's trade. The tin contained blasting caps while the tool is a crimping and cutting tool for setting the caps in the sticks of dynamite. Collection of Ron Bonmarito.

An unusual and very fine miner's tool, this is a plumb bob for use in surveying the inside of a mine. The top of the brass tool is actually a wick lamp which burned oil and gave off enough light that the mining engineer could see it in the dark of the mine tunnels. Collection of Ron Bonmarito.

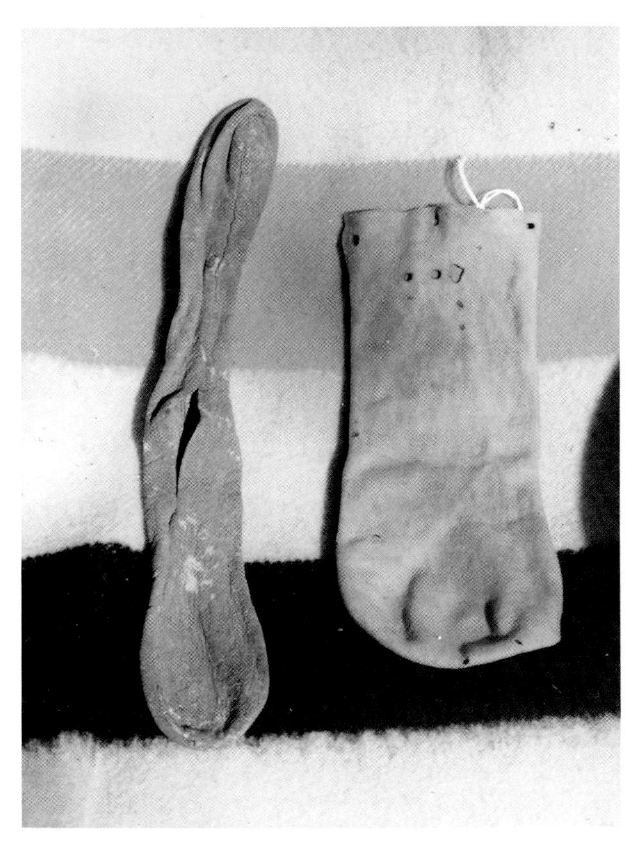

Two early gold pokes made from what looks like deer skin. The one on the right is rather conventional in style but the one on the left has an opening in the center of one side so it could be filled and the middle held together. Collection of Ron Bonmarito.

A rather short piece of drill steel for use in starting a hole when hand drilling a blasting hole in the ore. The hammer is actually a miner's pick or prospector's pick which has a small brass plack with the owner's initials engraved on it. The handle is cut for a loop which hung from the owner's belt. Collection of Ron Bonmarito.

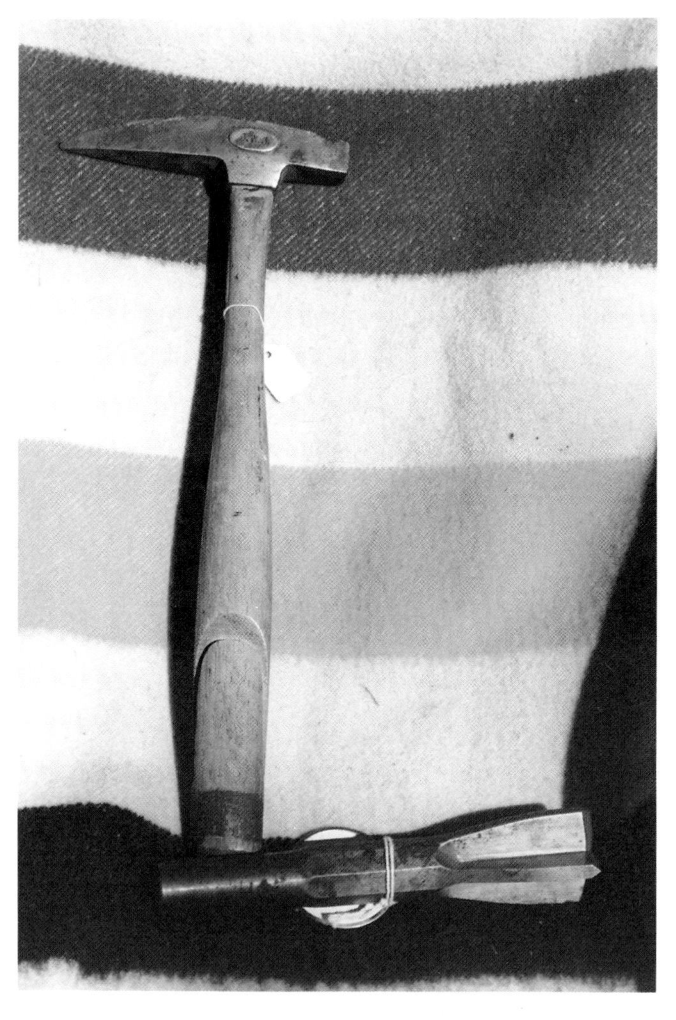

Three signs likely to be found in one form or other in every mine. These are all porcelainized metal signs designed to last a long time under adverse conditions. Collection of Ron Bonmarito.

An original miner's 'hard hat', this well preserved specimen is made from leather and some sort of fabric with the label that it is 'hard boiled'. The carbide lamp is a nicely made type of what appears to be German Silver, actually a nickel compound. Collection of Ron Bonmarito.

The ultimate mining collectible this restored cart rests in a local front yard. The wooden stick is 4 feet long and shown for scale. Note that the base is made of built up wooden pieces. Author's Collection.

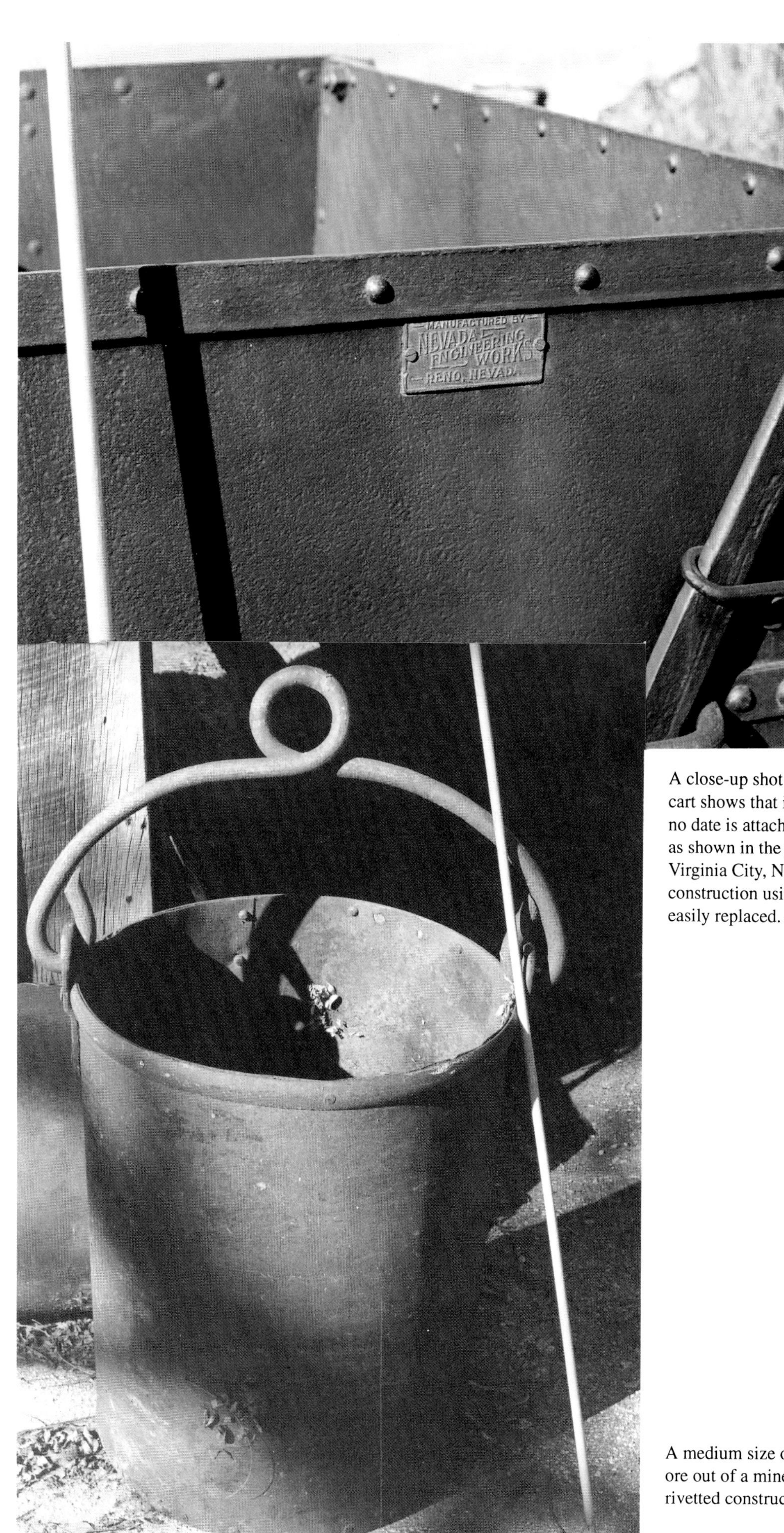

A close-up shot of the makers plate on the above ore cart shows that it was made in Reno, Nevada although no date is attached. This cart is exactly the same design as shown in the photograph of the Virginia Foundry in Virginia City, Nevada. This also shows clearly the construction using rivets and sheet steel which could be easily replaced. Author's Collection.

A medium size ore bucket used for hauling up rock or ore out of a mine shaft it is cylindrical in shape and rivetted construction. Collection of Ron Bonmarito.

Another type of ore cart utilizing solid wheels and a heavy iron carriage. The one has seen much use. Collection of Ron Bonmarito.

A very large ore bucket entirely of rivetted construction. This one is definitely large enough for a man to stand in as it is almost 4 feet high. Collection of Ron Bonmarito.

An unusual collectible, this is a "blowpipe kit" for doing sample analysis of gold ore samples. Old time chemistry students will understand how this works and it was an essential part of prospecting and mining. Collection of Ron Bonmarito.

No. 438 Series M 4695

MEMBERSHIP CARD

THIS is to certify that at the date of issuing this card APR -6 1912
Mr. Herman T Radke is a member in good standing
of THE MINERS' PROTECTIVE ASSOCIATION.

Signature of member
Herman T Radke

Secretary

Age 49
Height 5-6
Color Eyes blue
Color Hair Light
Weight 150
Complexion medium
Nationality German

FOR THE PROTECTION OF HOLDER

NOT TRANSFERABLE

This card is the exclusive property of THE MINERS PROTECTIVE ASSOCIATION, who may recover and take possession of it at any time. It cannot be sold or loaned, and if presented by anyone but the original holder will be taken up and CANCELLED. If lost notify the Secretary at once.

The above conditions are accepted.

(Sig.) H T Radke

A miners membership card in the Miners Protective Association dated April 6, 1912 for one Herman Radke. No indication where this was from but by this time many of the mining areas had unions which were more or less effective depending on the owners and the conditions in the mines. Collection of Ron Bonmarito.

A sampling of colorful miners union ribbons. Clockwise these are from Eureka, Nevada, Grass Valley, California, Park City, Utah, Butte, Montana, and Butte, Montana. Collection of Ron Bonmarito.

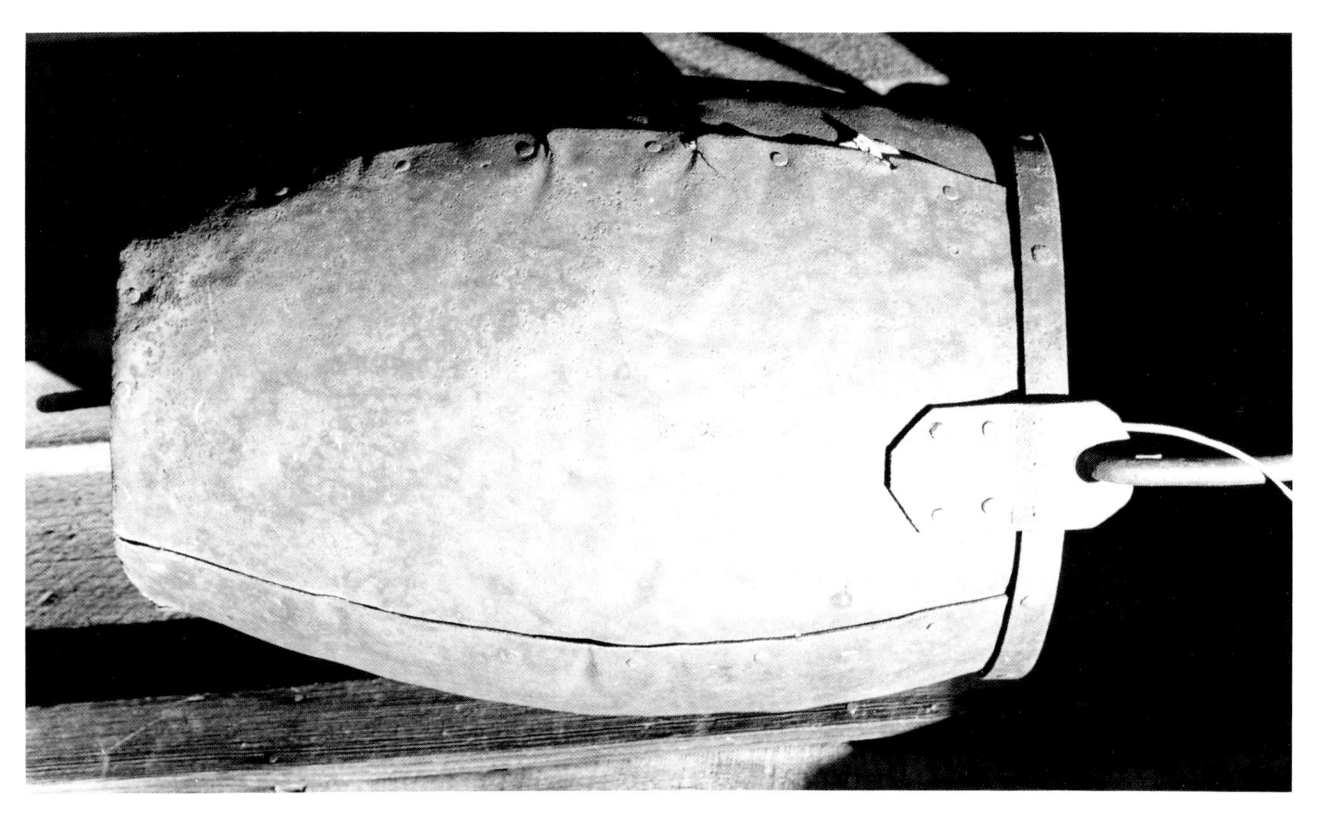

A crudely made ore bucket this one shows evidence of having been made from scrap sheet iron hammered and rivetted together. Apparently it worked all right. Collection of Ron Bonmarito.

A cross section view of lead silver ingot molds, the bottom three are from Eureka, Nevada while the top one is from Owens Lake, California and would have been used in the Cerro Gordo Lode. Collection of Ron Bonmarito.

The holder on the left is of exceptionally ornate design while the one on the right uses a different approach to holding the candle. Collection of Ron Bonmarito.

These three candle stick holders are all folding pieces, factory made and finely finished. Folders are fairly rare and could be carried in your pocket like a folding knife. Collection of Ron Bonmarito.

A good example of a miners lunch pail this one is in almost pristine condition. Collection of Ron Bonmarito.

Another ore cart of Virginia City Foundry manufacture, this one could hold about 1800 pounds of rock. Collection of Ron Bonmarito.

On a more practical level are the simple tools that didn't wear out for the average miner, including illumination methods. The earliest method of bringing light underground was using torches, but these were quickly replaced by candles. Candles do not give off much light and when set down on the floor of a mine they tend to get in the way, so the need for a a way to bring light up to the work level without snuffing it out led to the development of the miner's candlestick. Miners' candlesticks were easy to make for the average mine-employed blacksmith, and they allowed the standarization of the size of candles purchased for the miners. Candlesticks have a lot of variety to them, and can be found to a greater or lesser degree throughout the mining areas. They can bring rather serious prices for the collector. Condition and rarity are the determining factors of value. Virtually all candlesticks did get used, and consequently few will be found in new condition.

Transition pieces from candle stick holders to oil lamps. The candle stick holder has a hollow handle for holding matches in a waterproof container while the oil lamp is rather crudely made and poorly soldered together. Collection of Ron Bonmarito.

The top miners candle stick holder is finely made while the bottom one has been adapted to hold an oil lamp. Collection of Ron Bonmarito.

Next in line of the illumination devices is the oil lamp. The oil lamp was a simple device intended to be worn on the head or on a hard hat, secured by a tie or some other method. It was simply a small metal reservoir with a wick which came out the top and was usually made of brass. Oil lamps are fairly rare today in this country because they were less popular than the carbide lamps which were developed about the same time. Oil lamps did not usually have a reflector to bounce the light, since reflectors tended to soot up rather quickly. They also tended to be a bit dirtier than candles, which were easier to use.

While the candlestick and oil lamp were in use in North America, European miners brought over English candlestick lamps. These were artistically designed brass lamps made to hold a candle, with a grate around the flame to prevent ignition of coal gas (methane). The design proved effective because it allowed small explosions to occur on the grate from the contact of the flame and the gas, preventing a disastrously big explosion. Since the amount of light given off by these lamps was somewhat less than from a bare candle, and since Western gold mines did not have a major problem with methane, miners in the U.S. did not feel compelled to use them. Still, there is a bewilderingly large variety of these lamps to be seen. In some areas of the world they are still being produced.

This is an excellent example of a "General Grant" lamp, a candle lamp popularized by U. S. Grant when he visited the Virginia City Nevada mines. Collection of Ron Bonmarito.

Crew photograph of a Comstock mine. The man in center, bottom row is holding an English miner's lamp, unusual for the Comstock. He probably brought it with him from the old country. Collection of Ron Bonmarito.

Next in line of development was the carbide lamp. This simple device consisted of a jar-type container made of brass, with a top that screwed on to the base. The top part held a nozzle, or 'nipple', with a shiny reflector to bounce the light produced by a flame coming out of the nipple. This flame was ignited by a flint striker attached to the nipple and the whole thing worked when water was added to the carbide (actually calcium carbide) in the bottom container. The meeting of the two chemicals produced the highly flamable gas acetylene, which burns with a very clean bright blue flame. When this lamp is kept clean, it produces a very usable and very bright light. Its only drawback is that the miner must replenish the materials periodically, with the frequency depending on the size of the reservoir and how well he has adjusted his lamp. These lamps were produced in huge numbers from the 1890s on and can be found in many antique shops. The collector should be aware that they are still being produced and used in less developed parts of the world.

A nicely made carbide lamp dating from the turn of the century this one has attachments on the back for a hat or hand hold. This one is made of plated brass or german silver which actually is a nickel alloy. Collection of Ron Bonmarito.

A larger style carbide lamp made to hang off a timber or rock this lamp is in like new condition. Collection of Ron Bonmarito.

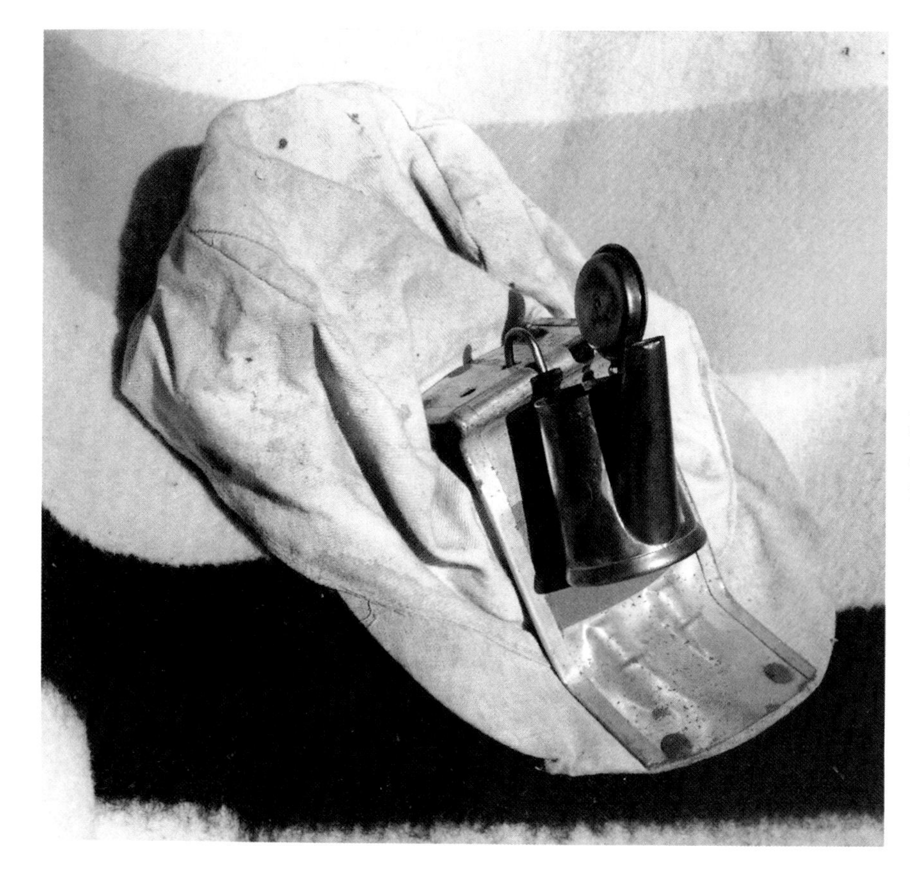

An early cloth miners hat with a wick lamp attached. The wick lamp has the top open but no wick in place. The hat itself is simply made of cotton. Collection of Ron Bonmarito.

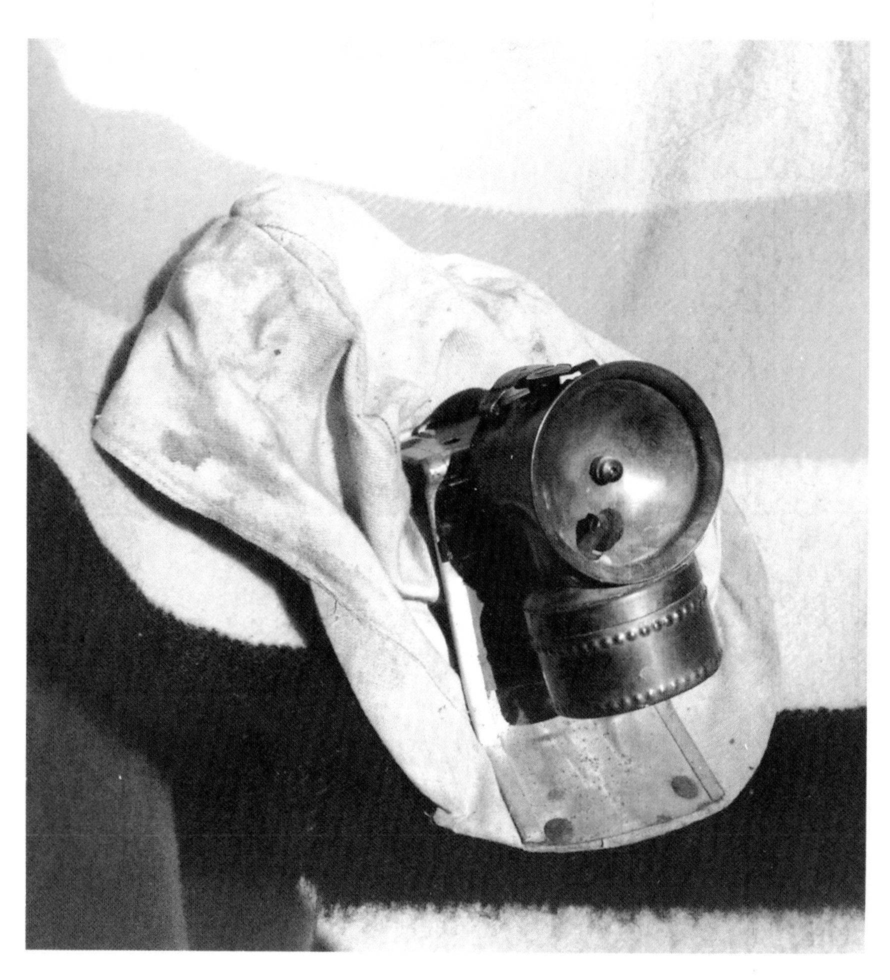

The same hat as shown above with an early brass carbide miners lamp attached. The hats were made to accept a variety of attachment styles. Collection of Ron Bonmarito.

Almost all placer miner had access to a set of gold scales, if they didn't own sets themselves. They were an essential item, for they safeguarded against cheating by unscrupulous buyers and shopkeepers. Since real minted gold and silver coins were scarce in the early gold fields, the accepted medium of exchange was gold dust. No matter that it might be adulterated with a little black sand, or that the gold was only 75% pure; it was gold, and everybody expected fair value for a correct weight. Because these were such necessary tools for everyone involved with business in a mining town, there are many small gold scales to be found in antique stores. Prices vary enormously according to condition and the quality of manufacture. These scales were made all over the world (indeed, they continue to made today) and the truly fine sets are worth a lot of money. These scales are true balance scales. The weights that accompanied them have frequently been lost because of their small size. They are great items for displaying and are very popular with antique enthusiasts without a particular interest in mining. Weights were usually made of brass, which won't corrode and is easily machined.

Occasionally, adulterated weights can be found. The most common method of tampering was to hollow out the brass and fill the weights with lead. These will bring high prices to the right buyer.

Above right:
This is a small silver ingot mold holding 40 ounces of silver, not really very much, but it could be used for gold as well. Collection of Ron Bonmarito.

Two miners gold scales, the top one a bit higher in quality but missing some of the weights. Collection of Ron Bonmarito.

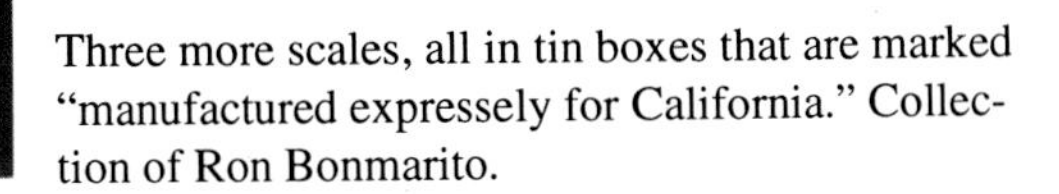

Three more scales, all in tin boxes that are marked "manufactured expressely for California." Collection of Ron Bonmarito.

CURRENCY

One of the real problems that existed in the gold fields, especially in the earliest days, was the distrust of paper money and the lack of silver coins in smaller denominations. When this problem reared its ugly head, a solution was quickly developed. They began to produce coins, or tokens, of brass or some other metal. Since, then as now, it was illegal to mint coins (only the U.S. government can do that), these tokens usually have an admonition on them that restricted their usage to a particular store, hotel, or brothel. This circumvented the law, but they were still bought and sold like money, traded for various goods, traded between cooperative shop owners, and generally used just like real money was. Today these tokens can frequently be found in the collections of coin dealers and traders and occasionally in antique shops. The rarity is for them to be found in a 'dig', but that also happens. Since they were usually made of brass, less often of aluminium or steel, their survival rate can be high. Prices for these artifacts of the era occasionally bring premium prices. They are much sought after today by collectors.

Paper money of the day was generally not considered valid money, especially in the actual gold camps and boom towns. Records exist of paper dollars being exchanged for as little as 50% of face value in exchange for food and materials. Real early paper money is truly scarce today, about on the level of real Confederate paper dollars. Determining their value is relatively easy, since there are established factors which take into account condition and rarity.

There is a wide variety of miner's hardware that can be collected. Among other items are things like bottles, lunch buckets, paychecks, timesheets, and ore assays.

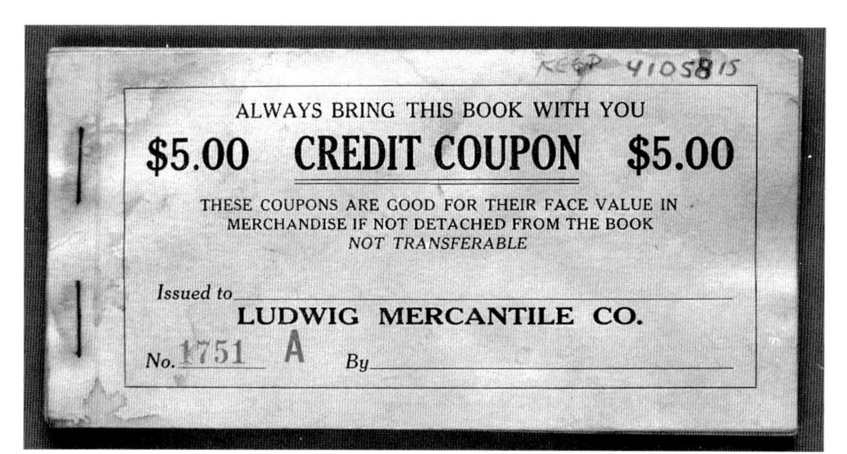

One of the many coupon books used by local and company stores in the western mines. This one from the town of Ludwig, Nevada contains a number of smaller denomination coupons which could only be used in that particular store. In the absence of small change this was what merchants gave out. Collection of Ron Bonmarito.

A thinly disguised attempt at providing currency in the absence of small denomination U.S. bills. This one at $3.00 is not likely to be mistaken for a regular bill.

Another of the Manhattan Silver Mining Company's artificial money. These were probably illegal but they were used nonetheless in an area where silver was king.

MINING STOCK CERTIFICATES

Today's general public cannot really understand the fervor of gold investors during this Gold Rush era. To day we are accustomed to "stock markets" quoted by the nightly news and subject to more government regulations than any one person can comprehend. During the mid-1800s and later no such regulation existed, other than perhaps a requirement that a corporation issuing stock register with the state where it was doing business. At its most fundamental level, stock in a corporation could be purchased by asking for a certificate which was filled in by hand by an authorized officer of the corporation. The certificate was numbered by hand and written down in a bound book kept for such purposes. If you had a brand new mine and wished to raise capital to develop it, it was almost that easy—assuming that you could find someone to buy the stock and that you could still retain control over your own corporation.

Stock certificates were bought and sold in the open streets in places like San Francisco and Sacramento, where smart promoters could manipulate prices to make enormous profits within days, or sometime hours. Even the smallest of mining districts quickly developed places for the exchange of stocks on something of an orderly basis. Employees of the mines were frequently paid in part with shares in the mines where they worked, which they were happy to get as long as the ore coming out of the ground was high-grade. It was a fairly common ploy of mine owners to offer their employees stock at a price slightly above the issued price when things were going good. The employees, knowing that the mine was profitable, were reluctant to sell their shares because they could see the future price increasing to the point where they could make some real money. On the other hand they needed cash to live with so they frequently sold the shares at bargain prices (or so they thought).

The next thing to happen was the increase in price brought on by greed, not by the true value. This resulted in an artificially inflated market price. At this time, the company owners would quietly sell out most of their shares, using shills and relatives to disguise their true actions. Then the mine owners would circulate a rumor that the mine had run into some sort of problem, such as a pinched out vein, an unexpected flood of water, or a cave-in that closed a particularly good shaft. The owners, being smart (and rich), would take into their confidence a writer, usually a newspaperman, who would be invited to a 'confidential' tour of the facilities. He would then write an article detailing the problems of the mine and the public would react by selling their shares of stock at artifically low prices. The mine owners, having already sold their stock shares at higher prices, would then buy up the now cheaper shares and achieve complete control over the mining corporation. The corporation would then declare some exhorbitant dividends payable to owners of record after the manipulation, and reap the profits.

This was an all too common event in the boom towns of the western mining camps, with the Hearst family being past masters of it. They used this type of ploy to achieve eventual control over the Homestake mine in South Dakota. They were by no means the only ones to do this, of course, and there are many stories about the frantic buying and selling of stocks in these locales.

What is of interest to us now is the existence of these mining certificates as collectors items. While they are of no value as stock certificates (the corporations no longer exist in most cases), they do have substantial value to collectors. The earliest of certificates were plainly printed sheets of paper and resemble large checks. They were common at one point in time but have largely disappeared from public circulation except in collections and antique shops. Later on the certificates were produced from engraved plates. These paper items are true works of art.

It is not within the scope of this book to delve into the valuation of such items, but it is worth noting that considerable value is involved in these certificates. It is most important here to note the interest shown in the stocks of particular mines. Especially interesting are those for mines in Virginia City, Nevada, because virtually all of them were under corporate ownership and their shares were publicly traded with great vigor in San Francisco and New York. They were also very successful over many years. Some paid handsome dividends before going out of production.

Part of the interest in the Virginia City Comstock lode mines stems from the relatively easy ride from the San Francisco Bay area to Virginia City, Nevada, all by train. In the heyday of train travel it was possible to get on a Union Pacific (UP) train in San Francisco in the morning and by evening be in Virginia City, arriving on the reliable Virginia and Truckee (V & T) which connected with the UP in Reno. After the V & T was opened in 1869 most mine-owners were located in San Francisco, and the whole system depended on hired help to keep things running. This was a far cry from the mines in Austin, Eureka, or Ruth, which were a long ride from the nearest railhead at best.

Because of fast transportation methods, the Virginia City mines were operating in plain view of everybody, and as a result great interest was paid to them. Information was easy to come by; it was necessary only to walk into any saloon in Viginia City and listen. All the latest rumors were there for the asking and the papers were full of the happenings of the mines.

This cartoon was originally published in a newspaper or periodical in 1864. It depicts in a most unflatering manner certain well-known (then) stock sellers. They were known as 'bummers' in reference to the practice of selling worthless mining stocks. The original is quite small, about 1 inch by 2.25 inches, and very fragile. Collection of Ron Bonmarito.

ESMERALDA DISTRICT.

SILVER EEL GOLD & SILVER MINING CO.

800 SHARES

$100 EACH.

CAPITAL STOCK.

$80,000.

INCORPORATED JANUARY 24th, 1863.

Aurora, Mono Co. February 14 186

This Certifies, That A. A. Bermudez is entitled to Fifteen ... Shares in the Capital Stock of the SILVER EEL GOLD AND SILVER MINING CO.

Transferable on the Books of the Company, by endorsement hereon and surrender of this Certificate.

Secretary. President.

A stock certificate from a mine in Aurora, Mono County California in 1863. Actually Aurora was in Nevada, they just didn't know it yet. This stock is particularly ornate for this time period. This certificate would be worth something over $800 to a collector today. Collection of Ron Bonmarito.

INCORPORATED UNDER THE LAWS OF THE STATE OF NEW YORK.

No. 198 This is to Certify, that

entitled to — One Hundred — Shares

100 SHARES

OF THE CAPITAL STOCK OF THE

BLACKFOOT GOLD MINING COMPANY

LOCATION OF MINE, ONEIDA COUNTY, IDAHO.

transferable only on the Books of the Company in person or by Attorney, on surrender of this Certificate

New York, March 1st 1882

Secretary President

100,000 SHARES

SHARES $10 EACH.

MUTUAL TRUST COMPANY

A stock certificate from a mine in Idaho, appropriately the Blackfoot Gold Mining Company. The interesting thing to note here is that the company was incorporated in New York. This was not atypical of western mining companies, which drew big money from both coasts. Collection of Ron Bonmarito.

This certificate is dated 1881, San Francisco, for a mine in Esmeralda County Nevada. Actual size is not much bigger than a typical business check of today. Collection of Ron Bonmarito.

From Ione City, Nevada, this certificate is dated August 25, 1863, and was actually for a company that was going to mine silver and gold, not copper. Collection of Ron Bonmarito.

An early certificate from Gold Hill, Nevada on December 28th 1863. This Gold Hill Silver and Mining Company certificate is ‘unassessable stock’, which means that the mine owners could not require the stockholder to put up more money for each share of stock. Some mines actually never produced any bullion and continually assessed their stockholders for more and more money, finally going bankrupt. Collection of Ron Bonmarito.

INCORPORATED UNDER THE LAWS OF THE STATE OF COLORADO.
CAPITAL $ 3,000,000. THE SHARES $10. EACH.
81
DENVER MOUNTAIN MINING COMPANY OF DENVER, COLORADO.
This certifies that O. Conover
is the owner of Five Hundred Shares of the Capital Stock of THE DENVER MOUNTAIN MINING CO. of the par value of TEN DOLLARS each, fully paid and non assessable transferable by endorsement and delivery, and on the Books of the Company in person or by Attorney and on surrender of this Certificate.
In Witness Whereof the Seal of the Company is hereby affixed together with the signatures of the President and Secretary, at the City of Denver this day of 1882
O. Conover SECRETARY PRESIDENT
300,000 SHARES

Dated 1882 this certificate from Denver, Colorado is much more ornate than earlier types. Collection of Ron Bonmarito.

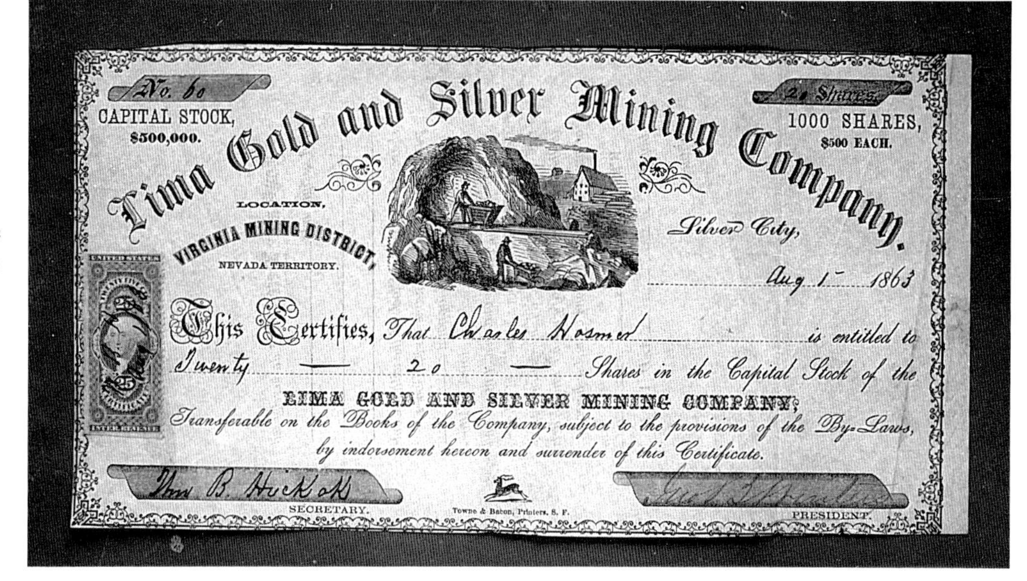
No. 60
20 Shares
CAPITAL STOCK, $500,000.
1000 SHARES, $500 EACH.
Lima Gold and Silver Mining Company.
LOCATION, VIRGINIA MINING DISTRICT, NEVADA TERRITORY.
Silver City, Aug 1st 1863
This Certifies, That Charles Hosmer is entitled to Twenty — 20 — Shares in the Capital Stock of the LIMA GOLD AND SILVER MINING COMPANY; Transferable on the Books of the Company, subject to the provisions of the By-Laws, by indorsement hereon and surrender of this Certificate.
Wm B. Hickok SECRETARY. PRESIDENT.

This Lima Gold and Silver certificate from Silver City, down the road from Gold Hill and Virginia City, Nevada, is dated August 1, 1863—before Nevada was a state. Collection of Ron Bonmarito.

CAPITAL STOCK $ 480.000
No 210
20
Birdsall Mining Company.
MONTOUR LEDGE
INCORPORATED MARCH 11th 1863.
VIRGINIA CITY, N.T. August 24th 1863
THIS CERTIFIES that A. C. Hunter is entitled to Twenty Share in the Capital Stock of the Birdsall Mining Company. Transferable on the Books of the Company upon the surrender of this Certificate properly endorsed
1600 SHARES
1600 FEET

Another certificate from the Territorial days of Nevada. This one from the Birdsall Mining Company is interesting because it claims 1600 shares for 1600 feet of the lode. This was typical in the earliest days of mining in the West and related back to the practice of a 'share' being one foot of ground in a claim. Collection of Ron Bonmarito.

A certificate from one of the richest mines in the Gold Hill/Virginia City district, the Yellow Jacket Silver Mining Company. Shares were valued at $1,000 par value. Not exactly a cheap investment then or now, also shown at one share for one foot. Collection of Ron Bonmarito.

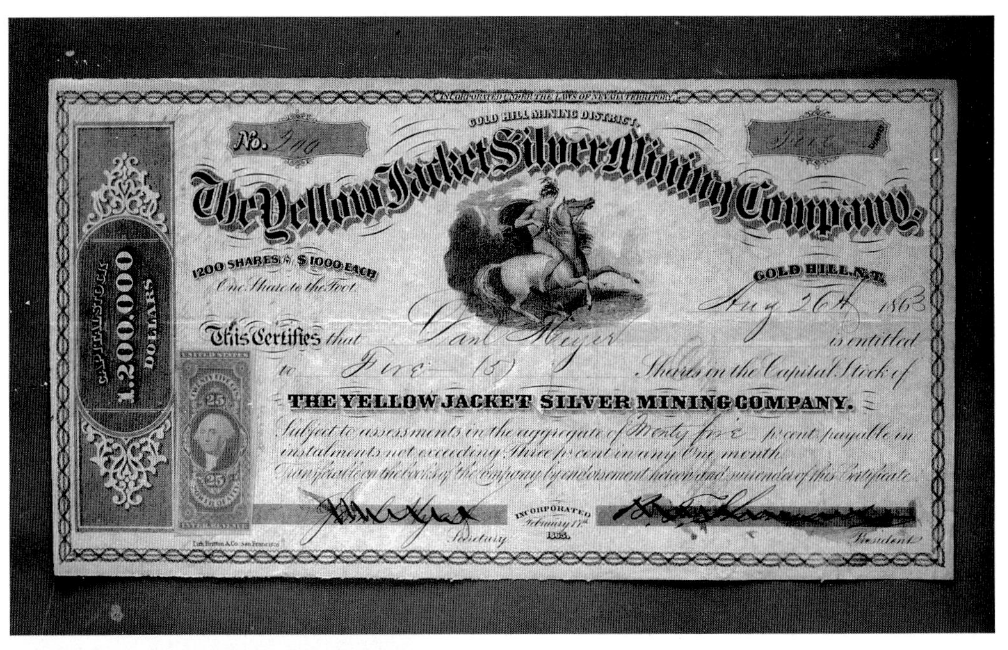

This printed certificate from the Antelope Gold and Silver Mining Company is dated in 1880. It is about the size of a standard check of today. Collection of Ron Bonmarito.

Colorado mines seem to have had better access to good printing presses. This certificate is very fancy, yet is dated 1867. Incidently, a capital of $400,000 is not particularly large. Collection of Ron Bonmarito.

Another certificate from the Nevada Territory, this one is from Devil's Gate District, which is part of the Comstock lode. It also shows the description of "one foot to one share." Collection of Ron Bonmarito.

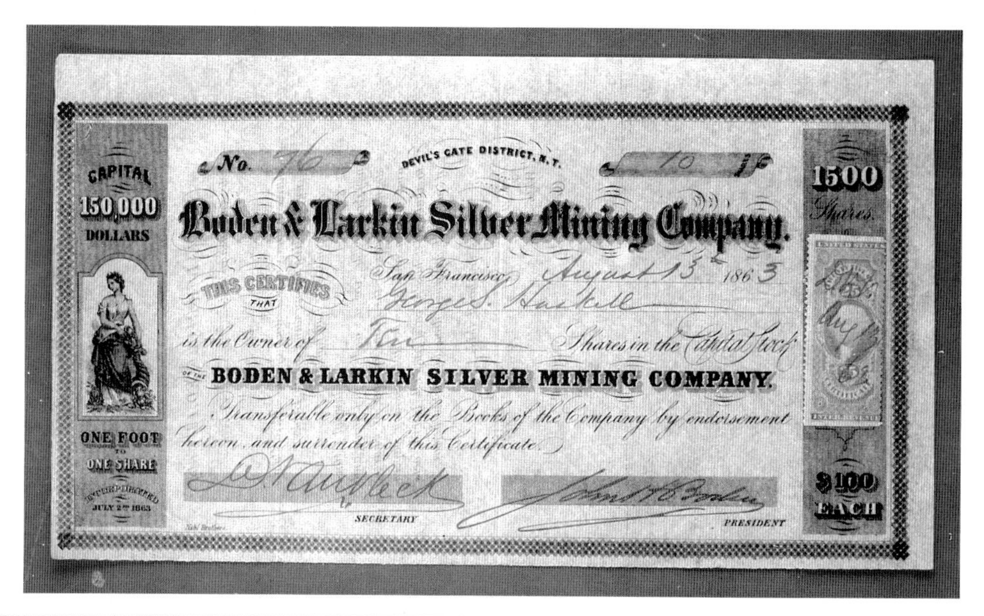

Some mines had imaginative names. This one was called the "Live Yankee Gold and Silver Mining Company." It is not known why the name was chosen, but the mine was in the Reese River District which included Austin, Nevada. Collection of Ron Bonmarito.

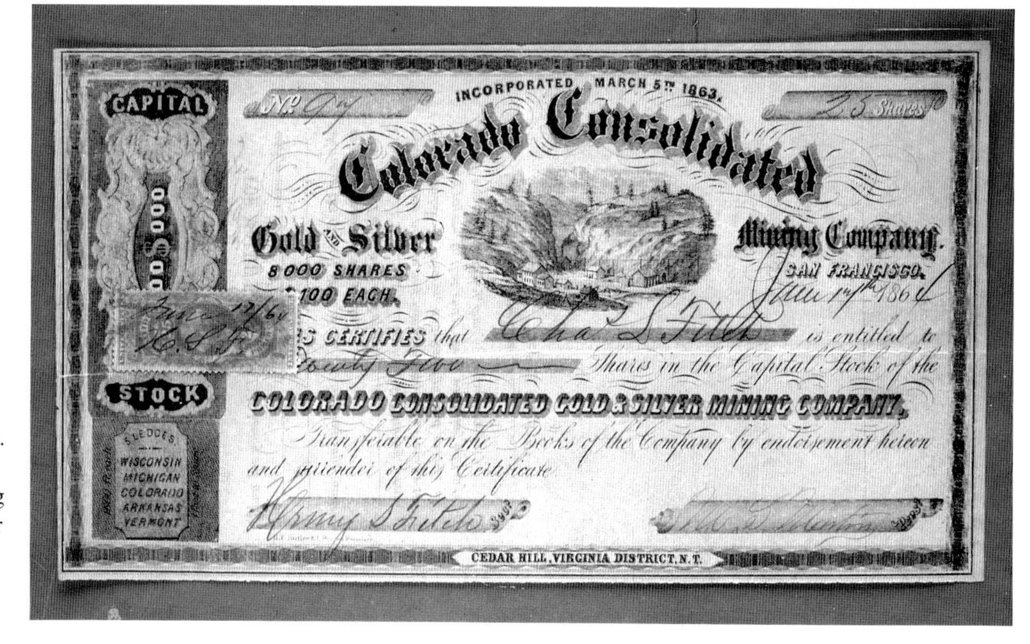

You couldn't always depend on the name to indicate where the mine was. This certificate from the Colorado Consolidated Gold and Silver Mining Company is actually located at Cedar Hill, Virginia District, Nevada Territory. Collection of Ron Bonmarito.

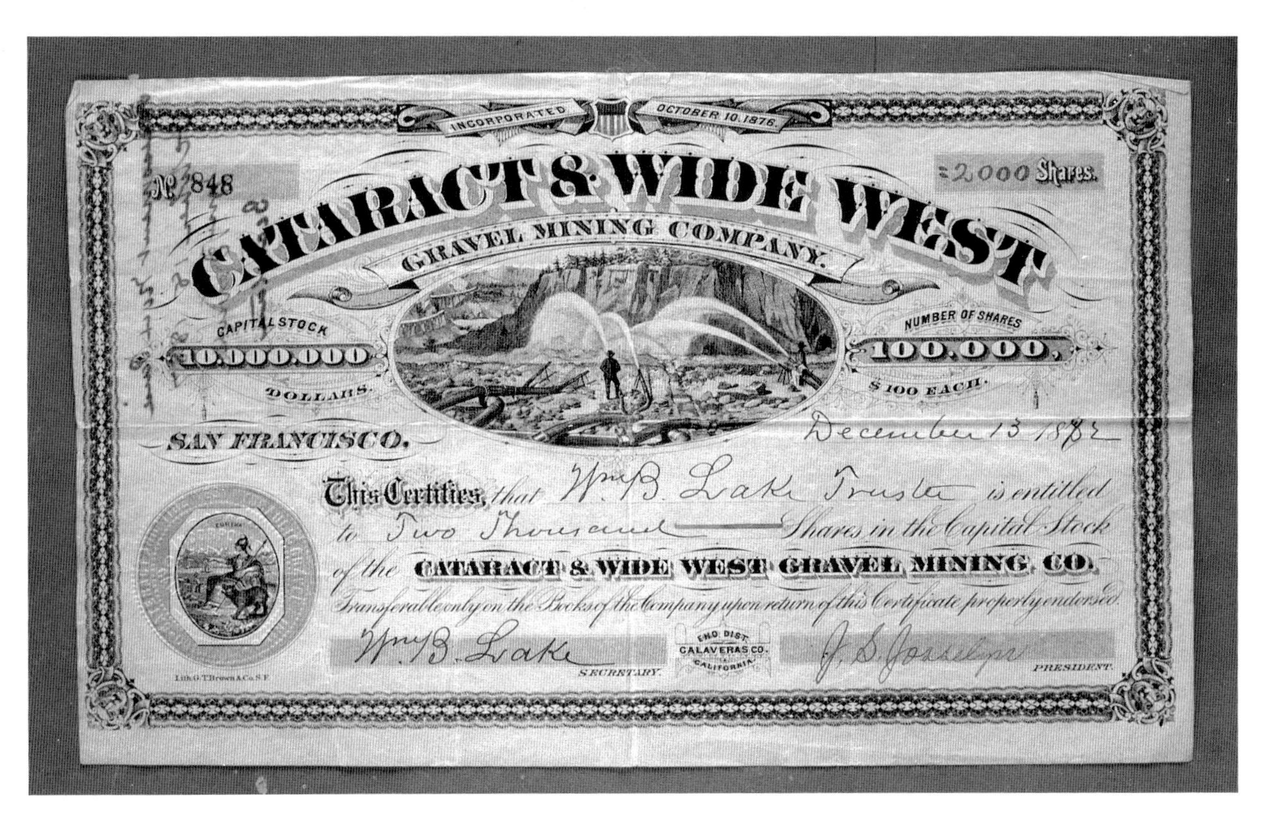

INCORPORATED OCTOBER 10, 1876.

No. 848

2,000 Shares.

CATARACT & WIDE WEST

GRAVEL MINING COMPANY.

CAPITAL STOCK 10,000,000 DOLLARS.

NUMBER OF SHARES 100,000. $100 EACH.

SAN FRANCISCO, December 13 1882

This Certifies, that Wm B. Lake Trustee is entitled to Two Thousand Shares in the Capital Stock of the CATARACT & WIDE WEST GRAVEL MINING CO.

Transferable only on the Books of the Company upon return of this Certificate properly endorsed.

Wm B. Lake SECRETARY.

2ND DIST. CALAVERAS CO. CALIFORNIA

PRESIDENT.

Nor could you depend on the description on the certificate to find out what the company mined. This one from the Cataract & Wide West Gravel Mining Company was not mining for gravel but gold. Incidently, with $10,000,000 in authorized stock, this was a heavily financed company in Calaveras County, California. Collection of Ron Bonmarito.

LORD BYRON

GOLD & SILVER MINING COMPANY

No. 67

SHARES

CAPITAL STOCK. $180,000.

INCORPORATED SEPTEMBER 26th, 1862

1,800 Shares, $100 Each.

Capital Stock, $180,000.

Aurora, Mono Co., Cal., 1863

This Certifies,

That is entitled to Shares in the Capital Stock of the LORD BYRON GOLD AND SILVER MINING COMPANY.

...ferable on the Books of the Company by endorsement hereon and surrender of this Certificate.

1,800 SHARES.

SECRETARY.

PRESIDENT.

Agnew & Deffebach print, 511 Sansome Street, S. F.

This Lord Byron Gold & Silver Mining Company certificate dates from before the knowledge that Aurora, Nevada was not in California. Collection of Ron Bonmarito.

THE NEWSPAPERS

The papers! The newspapers of the time were some of the most ill-written, untruthful pieces of trash ever foisted on a believing public. Yet miners who could read were hungry for anything to exercise their eyes on, and were more than willing to read aloud the events reported in the papers of the day. **The Territorial Enterprise** of Virginia City, Nevada was one such paper. Much that was written in the paper should have been taken with a grain of salt because it was pure fiction. The newspapermen of that day were often little better than shills for the mine owners. On the other hand, they were unafraid to blow the whistle on what they deemed to be a wrong, and to back up their call with a Colt revolver if that was necessary.

The infamous Samuel Clemens, a.k.a. Mark Twain, was a writer for the **Enterprise** in the early days, and was responsible for some of the most outrageous lies ever to be printed as fact. His publisher, John Goodman, was no better. It has been theorized that Twain actually left Virginia City because some of the lies he had perpetrated had finally come home to roost. More likely, he left town because of the outstanding bar bills he had accumulated in the numerous saloons. Incidently, it is probable that his pen name was developed during his days in Virginia City for his habit of calling out to a bartender to "mark twain"—in other words, to put two drinks on the tab for him. In any case, Clemens didn't like Virginia City, much preferring the weather of San Francisco; that is probably the real reason he left town.

Other boom towns were judged by how many saloons they had and how many newspapers they supported. A town was considered 'on the map' if it had one paper, and really a big place if it had two competing papers. The logistics of producing a newspaper, not normally given much thought in today's world, were formidable in the 1800s. First, all the papers of that day were printed on large, heavy, very cumbersome presses. The copy was hand-set in movable lead type. Moreover, newspapers required supplies such as ink and paper. In the rush to supply a new town with every implement needed for mining, and for feeding and corrupting a populace (with booze), ink and newsprint were way down the list of things to be freighted to the area. In a few rare instances newspapers were handwritten and sold to a clamoring public. It is not surprising that surviving copies of papers from what are now ghost towns are rare items to possess.

In addition to the sensational aspects of the newspaper business on the frontier, papers can give us a look into the past not afforded by any other means. The news of the time included all the good and the grisly, with great embellishment at times and with simple straightforward words at other times. That they accomplished what they did is testament to the fortitude of their owners, since all of these papers died from lack of revenue. Some lasted for years, some only for days, but they all had an impact.

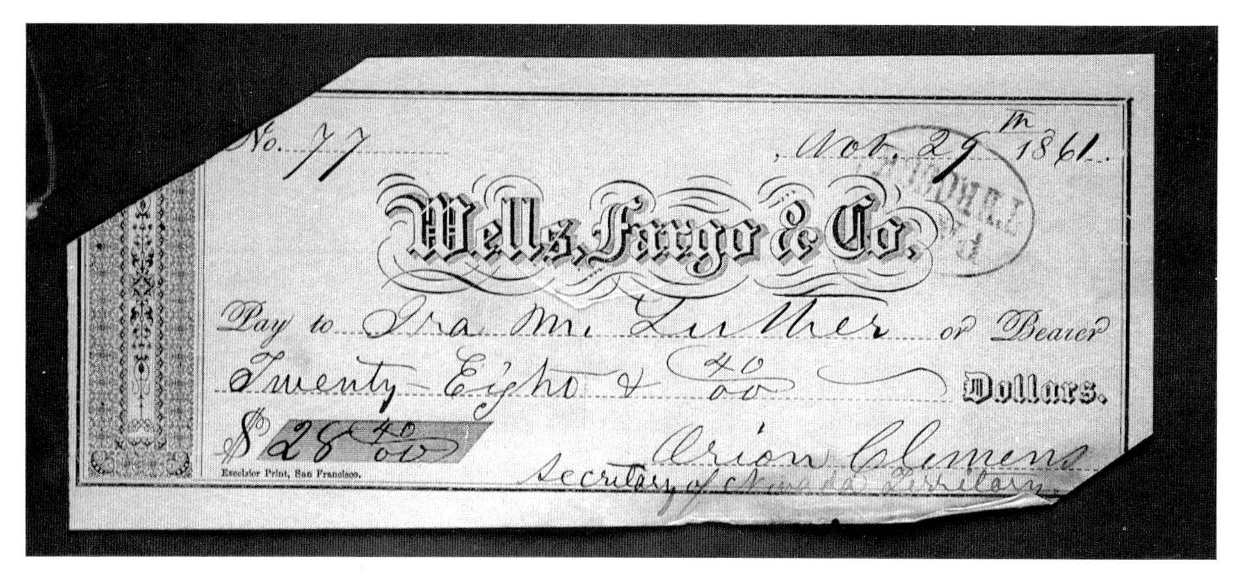
No. 77 Nov. 29th 1861.
Wells, Fargo & Co.
Pay to Ira M. Luther or Bearer
Twenty Eight & 40/00 Dollars.
$28 40/00
Orion Clemens
Secretary of Nevada Territory
Excelsior Print, San Francisco.

This receipt is only interesting because it was signed by Orion Clemens as Secretary of Nevada Territory. Mr. Clemens was the brother of Samuel Clemens, a.k.a. Mark Twain, and the one who got him a job in Virginia City. Virtually unknown today, Orion Clemens was a fine upstanding citizen when he got his ne'er-do-well brother a position at the Territorial Enterprise. Collection of Ron Bonmarito.

IN CONCLUSION

How big was the gold rush in the West? It was not the biggest as far as individual mines or mining districts go world-wide; the South African mines take that prize by a large margin. The Homestake Lode in South Dakota ranks about fourth overall in production; the Crippple Creek, Colorado District is next with the Comstock mines in Nevada ninth. The California Mother Lode ranks about twelth. Collectively, the gold and silver mines of the West (not including Alaska, which is another story altogether) ranks second only to the mines of South Africa.[22]

Despite the obvious objection that the numbers reported may not be accurate, the number of dollars produced from gold alone were indisputably huge. When silver production and later copper production are figured in, it isn't hard to see why mining in the western U.S. was an international phenomenon.

The search for gold continues in the western states, primarily in Nevada, where mountains of low grade ores exist which can be profitably mined only by utilizing a cyanide heap leaching process that has its liabilities. This is essentially a large earth-moving process, in which gold is removed from rocks in microscopic particles. There is nothing mysterious or romantic about this method of mining; it simply requires moving dirt from one spot to another and treating it with chemicals to recover the gold. The tailings are no more or less ugly than those of the hydraulics and dredges of the century before.

The past is frequently romanticized, but in reality it was anything but romantic, glamorous, or fun. The silver screen is full of images that we now consider the "reality" of mining in the West. But the next time you watch a movie about the Gold Rush, pull out this book. Take a look at some of the photographs of these miners, and of the people around them, and remember how it really was.

ENDNOTES

(1) Clarence A. Logan, *Mother Lode Gold Belt of California,* 43.
(2) Eliot Lord, *Comstock Mining and Miners,* 204.
(3) Lord, 205.
(4) Effie O. Read, *White Pine Lang Syne: A True History of White Pine County, Nevada,* 14.
(5) Read, 85.
(6) Read, 85.
(7) Read, 84.
(8) A. von Steinwehr, *Centennial Gazetteer of the United States,* 29.
(9) Jay A. Carpenter, *The History of Fifty Years of Mining at Tonopah, 1900-1950,* 23.
(10) Ovando J. Hollister, *The Mines of Colorado,* 11.
(11) Hollister, 134.
(12) Arthur Lakes, *Geology of Western Ore Deposits,* 120.
(13) Howard N. and Lucille L. Sloane, *A Pictorial History of American Mining,* 258.
(14) Joseph H. Cash, *Working the Homestake,* 12.
(15) Cash, 14.
(16) Cash, 15.
(17) Sloane, 153.
(18) Dorothy M. Johnson, *The Bloody Bozeman,* 7.
(19) Johnson, 85.
(20) Sloane, 228.
(21) Warren George, *Geologic Atlas of the United States Tintic Special Folio, Tintic, Utah,* United States Geologic Survey from 1900, 4.
(22) Vardis Fisher and Opal Laurel Holmes, *Gold Rushes and Mining Camps of the Early American West,* 26

SELECTED BIBLIOGRAPHY

Athearn, Robert G. *The Coloradans.* Albequerque, NM: University of New Mexico Press, 1976.

Brown, Robert L. *Central City and Gilpin County, Then and Now.* Caldwell, ID: Caxton Printers Ltd., 1994.

Brown, Robert L. *The Great Pikes Peak Gold Rush.* Caldwell, ID: Caxton Printers Ltd., 1985

Carlson, Helen S. *Nevada Place Names, A Geographical Dictionary.* Reno, NV: University of Nevada Press, 1974.

Carpenter, Jay A. "The History of Fifty Years of Mining at Tonopah, 1900-1950." *University of Nevada Bulletin, Geology and Mining Series #51,* Vol. XLVII, January 1953.

Cash, Joseph H. *Working the Homestake.* Ames, IA: Iowa State University Press, 1973.

Croft, Helen Downer. *The Downs, The Rockies and Desert Gold.* Caldwell, ID: Caxton Printers Ltd., 1961.

Egerton, Kearney. *Somewhere Out There, Arizona's Lost Mines and Vanished Treasures.* Glendale, AZ: Prickly Pear Press.

Emmons, David M. *The Butte Irish Class and Ethnicity in an American Mining Town,* 1875-1925. Urbana, IL: University of Illinois Press.

Fisher, Vardis and Holmes, Opal Laurel. *Gold Rushes and Mining Camps of the Early American West.* Caldwell, ID: Caxton Printers, Ltd., 1968.

Florin, Lambert. *Ghost Towns of the West.* Superior Publishing Co., Promontory Press, 1970.

Foreman, Bert. *Arizona Historical Land.* New York, NY: Alfred A. Knopf, 1982.

French, Richard. *Four Days from Fort Wingate: The Lost Adams Diggings.* Caldwell, ID: Caxton Printers Ltd., 1994.

Holister, Ovando J. *The Mines of Colorado.* 1867. Reprint, Springfield, MA: Samuel Bowles & Co., 1974.

Johnson, Dorothy M. *The Bloody Bozeman.* New York, NY: McGraw-Hill Book Co., 1971.

Kalt, William D. Jr., *Awake the Copper Ghosts: The History of Banner Mining Company and the Treasure of Twin Buttes.* Banner Mining Co., 1968.

Lakes, Arthur. *Geology of Western Ore Deposits,* 2nd ed. Denver, CO: Kendrick Book and Stationery Co., Publishers, 1905.

Logan, Clarence A. *Mother Lode Gold Belt of California,* Bulletin No. 108. Sacramento, CA: State Division of Mines, San Francisco, 1935.

Lord, Eliot. *Comstock Mining and Miners.* 1883. Reprint, San Diego, CA: Howell-North Books, 1980.

Morse, Edgar W., ed. *Silver in the Golden State.* Oakland, CA: Oakland Museum History Department, 1986.

The Mother Lode Country, Geologic Guidebook Along Highway 49 Sierran Gold Belt, Bulletin No. 141. Sacramento, CA: The State of California's State Division of Mines, 1948.

Paher, Stanley W. *Nevada Ghost Towns and Mining Camps.* Las Vegas, NV: Nevada Publications, 1970.

Paul, Rodman Wilson. *Mining Frontiers of the Far West, 1848-1880.* New York, NY: Holt Rinehart & Winston, 1963.

Potter, Miles F. *Oregon's Golden Years.* Caldwell, ID: The Caxton Printers, 1976.

Powell, Lawrence Clark. *Arizona, A History.* Nashville, TN: American Association for State and Local History, 1976.

Read, Effie O. *White Pine Lang Syne: a True History of White Pine County, Nevada.* Denver, CO: Big Mountain Press, 1965.

Sloane, Howard N. & Lucille L. *A Pictorial History of American Mining.* New York, NY: Crown Publishers, Inc., 1970.

von Steinwehr, A. *Centennial Gazetteer of the United States.* Philadephia, PA: Ziegler & McCurdy, 1873.

Stoll, William T. *Silver Strike: The True Story of Silver Mining in the Coeur d'Alenes.* Moscow, ID: University of Idaho Press.

Warren, George. *Geologic Atlas of the United States, Tintic Special Folio, Tintic, Utah.* United States Geologic Survey, Washington D.C., 1900.

Weis, Norman D. *Ghost Towns of the Northwest.* Caldwell, ID: Caxton Printers Ltd., 1971.

Wolff, Ernest. *Handbook for the Alaskan Prospector.* Fairbanks, AL: University of Alaska, 1969.

Wood, John V. *Railroads Through the Coeur d'Alenes.* Caldwell, ID: Caxton Printers, Ltd., 1984.

PRICE GUIDE

Price ranges quoted here are based on current selling prices or, in some cases, asking prices. Since most figures quoted are for western mining memorabilia in the western marketplace, they may be higher or lower than for other geographical markets. In some cases the rarity of the items pictured is such that current asking figures are the best estimates of the author and Mr. Bonmarito, who is an acknowledged expert in his field.

As in any collecting field, condition and rarity are the two most important governing factors affecting price in the market. It is not within the scope of this book to give definitive values to the collectables shown in this book. However, having said that, I feel the values quoted here are fair, objective and will serve to give the reader and collector a good indication of the worth of comparable items.

Page 137
Electric detonater this size $75
Crimping tool $25 to $50
Blasting cap tin $20
Printed postcard $50

Page 138
Plumb bob with case & etc. $1,000 in fine condition
Gold pokes $50 to $100—beware of fakes

Page 139
Prospector's pick, plain $35, fancy $250+
Drill, steel, hand-type, common $20
Sign, "surface" $150 to $200
Sign, "Chute" $150 to $200
Sign "Danger" $125 to $150

Page 140
"Hard Hat" $40 to $60 in fine condition
Lamp, $150 to $200
Ore cart $600 to $700 in fine condition

Page 141
Ore cart (as above) with local makers tag (add $150), so $800
Ore bucket, $225 to $260

Page 142
Ore cart $600 to $700
Large ore bucket, $325 to $375

Page 143
Assay kit, fire type (blowpipe) $300 if intact
Miners Union card $60 to $125 depending on person named

Page 144
Miners union ribbons, $75 to $350 depending on condition, locale and age.

Page 145
Ore bucket, crude $200
Lead-silver ingot molds, $100 to $550 for marked specimens

Page 146
Candle holder, left $200 to $300
Candle holder, right $500 to $700
Folding candle holders, $700 to $2,000 depending on manufacture and condition

Page 147
Miner's lunch pail, common $50 in fine condition
Ore cart, $400 to $600

Page 148
Oil lamp, $40 to $70
Candle holder with match case, $275
Candle holder with patent safety device $1,000
Oil lamp/candle holder, $200

Page 149
"General Grant" lamp, unmarked $400 to $500, marked $500 to $1,000

Page 150
Top carbide lamp, $200 to $275
Larger "eight hour" carbide lamp $75 to $110

Page 151
Cloth miner hats, $25 to $60 depending on condition

Page 152
Silver or gold ingot mold, smaller size $75 to $100
Miner's gold scales, center $250 to $300
Miner's gold scales, bottom $250 to $325 depending on completeness of weights and condition of tin.

Page 153 Currency:
Coupon book from Ludwig, Nevada–one of three known, and as such commands a premium price of $1,000. Condition is fine.
Manhattan Silver Mining Co script, unissued is $50 per certificate. If issued price goes up substantially!

Some generalities govern pricing of mining stock certificates and they need to be noted. First and most obvious is condition, second is age, third is the grade of paper stock used, and fourth is artistic quality of the certificate. In the west it is our experience that certificates from the Nevada mines of the pre-1880s will bring proportionately more than certificates from say, California of the same time. The exception to this rule is if the paper is from Bodie, California; these tend to bring premium prices.

Certificates signed by famous individuals such as Leland Stanford also tend to bring very high prices because of the notoriety of the signature.

Insofar as paper goes, those stocks printed on high-quality bank note paper stock will bring higher prices than those printed on plain paper. For Nevada, stocks issued before statehood will also bring higher values than issues done after statehood for the same corporation. Finally, artistic quality is somewhat subjective, but it becomes obvious that more artistically designed stocks are just better quality than those printed by the neighborhood printer.

Page 155
This first illustration is valued at $50 because of its rarity, not because it is a stock certificate.
Aurora, Nevada Cert. 1863 $750 to $900
Blackfoot Gold Mining Co., Idaho $100 to $140

Page 156
Esmeralda County, Nevada 1881 $130 to $150
Ione City, Nevada 1863 $400 to $475
Gold Hill, Nevada 1863 $450 to $500

Page 157
Denver, Colorado 1882 $125 to $150
Lima Gold & Silver, Silver City, Nevada 1863 $425 to $475
Birdsall Mining, Nevada 1863 $425 to $475

Page 158
Yellow Jacket Mine, Nevada $350 to $400
Antelope Gold & Silver Mining Co. 1880 $125 to $150
Colorado 1867 $175 to $225

Page 159
Devil's Gate District, pre-statehood Nevada $425 to $475
Live Yankee Mining Co. Austin, Nevada $450 to $500
Colorado Consolidated Mining Co., Nevada $425 to $475

Page 160
Cataract & Wide West Gravel Mining Co. in Calaveras, California $125 to $150.
Lord Byron G & S Mining Co. in Aurora, Nevada $450 to $500

Page 161
Paper receipt signed by Orion Clemens, brother of Samuel Clemens (a.k.a. Mark Twain).$300

INDEX